T0095380

SATELLITE BASICS FOR EVERYONE

C. ROBERT WELTI, PHD

SATELLITE BASICS FOR EVERYONE
AN ILLUSTRATED GUIDE TO SATELLITES FOR NON-TECHNICAL AND TECHNICAL PEOPLE

iUniverse books may be ordered through booksellers or by contacting:

iUniverse
1663 Liberty Drive
Bloomington, IN 47403
www.iuniverse.com
1-800-Authors (1-800-288-4677)

ISBN: 978-1-4759-2593-7 (sc)
ISBN: 978-1-4759-2594-4 (hc)
ISBN: 978-1-4759-2595-1 (e)

Library of Congress Control Number: 2012908983

Print information available on the last page.

iUniverse rev. date: 03/29/2018

Original illustrations by C. Robert Welti

Disclaimer

The information contained in this book is intended to be educational in nature and not for any other purpose. The author disclaims personal liability, directly or indirectly, for advice of information presented within. Although the author has prepared the manuscript with utmost care and diligence and has made every effort to ensure the accuracy and completeness of the information contained within, he assumes no responsibility for errors, inaccuracies, omissions, or inconsistencies. The Web is constantly changing and improving, so the author cannot guarantee that every reference will always remain constant.

Acknowledgement

Publication of this book would not have been possible without the help and support of several members of my family. My daughter, Sherri, who is a wonderfully talented professional graphic artist, designed the book cover. My wife, Kem, reviewed writing of the book and contributed many valuable comments and suggestions to enhance its presentation for clarity and understanding. She was my "north star" for making this book for everyone. My friend Suzanne Villarreal, who is a talented expository writing teacher, reviewed this book. She also gave me valuable information and advice on publishing this book in addition to being an ardent cheerleader. Numerous other family members and friends contributed many suggestions and encouragement. Thank you all!

Table of Contents

Preface

I was inspired to write this book after a presentation I gave to a group of retired men who have diverse backgrounds and work experiences, and know little about satellites. My talk included a hypothetical mission-planning task for placing a satellite in orbit and supporting it to perform its on-orbit mission. Response to the presentation was overwhelming, and many of the members wanted more information.

From the beginning, I realized that my presentation had to be simple, clear, interesting, and entertaining so I included abundant illustrations, familiar terminology, and simple explanations. For *Satellite Basics for Everyone*, I've developed even more of these materials, and hope to make satellite science as clear and understandable to you as it was for my audience of seniors, whatever your background, age, or work experience.

This book answers the following questions about satellites: What's a satellite? What are satellite missions? Who owns and operates satellites? Who builds satellites? How many satellites are in orbit? How close do they come to each other? How much debris is in orbit? What's the risk of collision? How do you launch satellites into space? How do you control satellites? How much do satellites cost? How do you put a satellite in orbit? What forces perturb the satellite? How do you keep a satellite in orbit? How do you keep the satellite oriented and pointing in the right direction? When is the best time to make orbit maneuvers? How much propellant do you need for corrections and maneuvers? What determines satellite lifetime? What's the future of satellite technology?

I wrote this book for grade school, high school, and college students, aerospace workers, and people who have curious minds. *The main objective of this book is to stimulate a broad interest in engineering and science.*

While looking at my book, my eight year old grand-daughter, Julia, asked, "Are they going to replace the Hubble Space Telescope?" (The book says its estimated lifetime is until 2014.) She also asked, "What does 'M' stand for?" I replied, "It stands for millions of dollars when used with the '$' sign." Then she exclaimed, "Wow! Satellites sure cost a lot of money." She continued to ask questions. My friend can't wait to get a copy of it for her thirteen year old grandson who is interested in science. Another friend who is involved in getting free books for grade school students wants my book. These examples show how it is a learning tool and a way to stimulate interest in satellites and science for grade school students.

As young students, my friends and I found it difficult to decide on career fields because we didn't have sufficient information. This book gives high school and college students an introduction to possible career field options.

In hindsight, development of satellites would have been easier and more efficient if all members of the satellite team had had a basic overall understanding of satellites and how they work. This reflection is especially true for the times when I worked on the international Intelsat V communications satellite and other satellite programs. Also, I noticed this same lack of overall satellite understanding when I managed satellite software development and maintenance for the U. S. Air Force Satellite Control Facility, and performed proposal evaluation and system engineering for U. S. Air Force satellite programs while at the Aerospace Corporation.

On one occasion while on an airplane, I was seated next to a United States Air Force Airman who was returning from a class on satellite intelligence resources. We discussed my book and he said, "I wish I had had your book before I took my class."

I showed my book to family and friends who knew little about satellites and how they worked. They said they learned a lot.

In summary, I wrote *Satellite Basics for Everyone* as clearly and understandably as possible for a wide audience. I hope you enjoy it.

C. Robert Welti

Introduction

An artificial satellite is an object which has been placed into orbit by human endeavor. Such objects are called artificial satellites to distinguish them from natural satellites such as the moon. This book gives an introduction and overview to artificial satellites that orbit the earth and how they are placed in orbit and kept there to perform their missions.

Each satellite has a use or a mission. With satellites we observe the universe and the earth, perform reconnaissance and navigation, make scientific measurements, and perform global communications. The satellite mission determines the satellite payload and orbit.

If the mission is to observe the universe, then the payload would be a telescope. If it's to provide communications, then the payload would be communications equipment. The rest of the satellite is the platform for supporting the payload. While transporting the payload around in its orbit, the platform provides electrical power, orientation, thermal control, and orbit control.

If you have ever wondered how fast satellites travel, how high above the earth they are, and how they move in space, this book is for you. I use familiar units of measure of miles for distance and miles per hour (mph) for speed. To make things easier, I only use a few fundamental equations with basic mathematical operations. But not to worry, all calculations are made for you and references provide derivations of equations. Appendix A includes conversion tables from metric to U.S. English units and vice versa that are helpful when looking up references or other literature research. Normalizing with respect to a surface circular satellite also makes calculations easier.

I start by summarizing the different types of satellites and the launch vehicles that boost them to orbit. I look at the number of operational satellites in orbit and the amount of space debris. I estimate satellite ownership and launch costs. Focus then turns to the geostationary communications satellite. The geostationary orbit is synchronous with the rotation of the earth, thereby keeping it on station at a specified longitude and constant altitude above the equator. Next, I describe typical communications satellite components. Then, in preparation for a mission planning example of putting a hypothetical geostationary communications satellite on station, I explain satellite motion.

Our mission planning example first transfers a stowed (folded) hypothetical communications satellite from a low altitude circular equatorial parking orbit into the geostationary orbit. I determine orbit parameters, velocity corrections, and orbit transfer times. Next, I refine mission planning by considering inclined orbits, perturbations, station-keeping, direct launch to orbit, and multiple transfer orbits. Then, I calculate the amount of propellant needed for the mission. Finally, I summarize the total mission mass.

I make observations and conclusions about how important satellites are to our lives, safety and security. I consider satellite technology advances, the best time to make orbit maneuvers, and lifetime limiting factors.

I define terminology throughout the book and provide a glossary of terms at the back of the book for easy reference. Look to "Rocket and Space Technology"[1] for derivations of equations used in this book. References and an index can be found at the back of the book.

After reading this book you may be able to impress your family and friends with your knowledge and understanding of an everyday integral part of our lives. If your thirst for knowledge is not satisfied with this book, you can follow the references as far as you like into the amazing world of satellites. It can be a stimulating, challenging and rewarding trip and I recommend "Rocket and Space Technology" to help you on your journey.

1

Artificial Earth Satellites

Satellite missions dictate their design and orbit. Satellites consist of a mission payload and a support platform or **bus**. The mission payloads perform their primary mission functions, including: making measurements, providing communications, providing navigation, and special military operations. The bus transports the mission payload around in its orbit and provides: electrical power, attitude control, payload pointing, temperature control, orbital position control, and orbital changes.

Origin of Artificial Satellites

The Soviet Union launched the first artificial satellite Sputnik 1 shown in Figure 1-1 on October 4, 1957. Sputnik 1 triggered the Space Race between the Soviet Union and the United States during the Cold War.

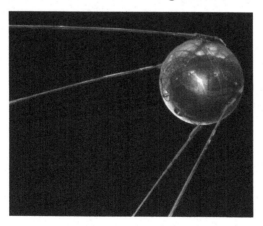

Figure 1-1 Sputnik 1 (NASA Public Domain)

In early 1945 the United States started the Vanguard Rocket development to launch a satellite. After several Vanguard explosions on the launch pad, as witnessed on TV, the United States launched its first artificial satellite, Explorer 1 shown in Figure 1-2 on January 31, 1958 with the Vanguard. The Explorer's scientific payload included instruments to measure: cosmic rays, temperatures, acoustics of cosmic dust impacts, and micrometeorite impacts.

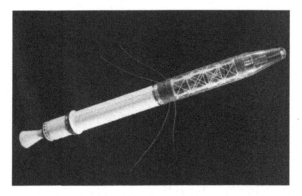

Figure 1-2 Explorer 1 (NASA Public Domain)

Major Satellite Types

Satellite missions dictate their design and orbit.

Major satellite types include:

- Astronomical
- Communications
- Navigation
- Reconnaissance
- Earth Observation
- Space Station
- Scientific

Astronomical satellites are satellites used for observing distant planets, galaxies, and other outer space objects. Payloads include telescopes and antennas and sensors sensitive to radiated energy at various wavelengths. The Hubble Space Telescope shown in Figure 1-3 is an example of a famous astronomical satellite.

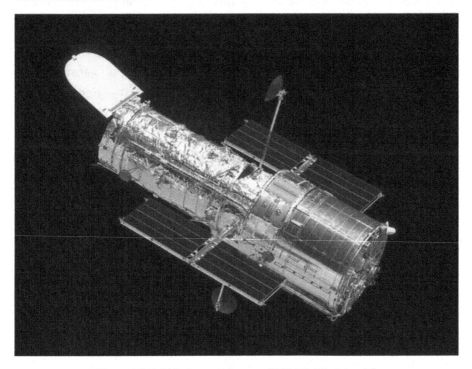

Figure 1-3 Hubble Space Telescope (NASA Public Domain)

The Hubble Space Telescope (HST) is a space telescope that was carried into orbit by a Space Shuttle in 1990 and remains in operation. A 7.9 ft. aperture telescope in low earth orbit, Hubble's four main instruments observe in the near ultraviolet, visible, and near infrared. Hubble's orbit outside the distortion of earth allows it to take extremely sharp images with almost no background light. The United States space agency NASA built HST with contributions from the European Space Agency. The Space Telescope Science Institute operates HST. Hubble's expected lifetime is until at least 2014.

Communications satellites are satellites stationed in space for the purpose of telecommunications. Modern communications satellites typically use 24 hour geostationary orbits, 12 hour Molniya orbits, or low earth orbits.

The most common orbit is the **geostationary orbit** where the satellite is placed on orbit above a point of the equator and orbits around the earth at the same rate as the rotation of the earth around its axis, thereby keeping the satellite on **station** above the equator.

Figure 1-4 shows a simplified drawing of a typical geostationary communications satellite. In Chapter 3, focus is on the *Geostationary Communications Satellite* because it affects our daily lives and our security.

Figure 1-4 Geostationary Communications Satellite

The Geostationary Communications Satellite provides global television, radio, business, Internet, and telephone services. Communications satellites are critical for national security because they provide global command and control of military resources and dissemination of vital intelligence information.

Navigational satellites are satellites which use radio time signals transmitted to enable mobile receivers on the ground to determine their exact location. The relatively clear line of sight between the satellites and receivers on the ground, combined with ever-improving electronics, allows satellite navigation systems to measure location to accuracies on the order of 10 feet in real time.

The Global Position Satellite (GPS) shown in Figure 1-5 is part of the constellation of satellites that provides global satellite signals for determining the position of receivers on the land, sea, and in the air.

Figure 1-5 GPS Satellite (NASA Public Domain)

GPS receivers also provide position information for missiles, lower earth orbiting satellites, drones and other weapon delivery systems. Position accuracy improves as the number of received satellite signals increases.

Reconnaissance satellites are earth observation satellites or communications satellites deployed for military or intelligence applications. Figure 1-6 shows an artist's concept of one of these satellites. Governments keep gathered information from these satellites classified.

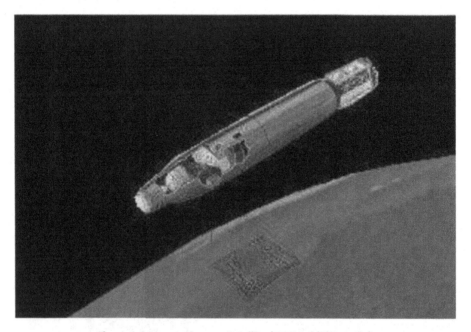

Figure 1-6 Reconnaissance Satellite (NRO Public Domain)

The first generation type reconnaissance satellites took photographs and ejected canisters of photographic film, which would descend to earth. Later satellites had digital imaging systems and downloaded the images via encrypted radio links. Examples of reconnaissance satellite missions include: high resolution photography (IMINT), measurement and signature intelligence (MASINT), communications eavesdropping (SIGINT), covert communications, monitoring of nuclear test ban compliance, and detection of missile launches.

ARTIFICIAL EARTH SATELLITES

Earth observation satellites are satellites intended for non-military uses such as environmental monitoring, meteorology, map making, and topology. They take measurements at various wavelengths.

The Geostationary Operational Environmental Satellite (GOES) shown in Figure 1-7 is operated by the United States National Environmental Satellite, Data, and Information Service (NESDIS) and supports weather forecasting, severe storm tracking, and meteorology research.

Figure 1-7 GOES Weather Satellite (NOAA Public Domain)

Spacecraft and ground-based elements work together to provide a continuous stream of environmental data. The National Weather Service (NWS) uses the GOES system for its United States weather monitoring and forecasting operations and scientific researchers use the data to better understand land, atmosphere, ocean, and climate interactions.

Space stations are man-made structures that are designed for human beings to live on in outer space. A space station is distinguished from other manned spacecraft by its lack of major propulsion or landing facilities. Space stations are designed for medium-term living in orbit, for periods of weeks, months, or even years. Figure 1-8 shows the International Space Station.

Figure 1-8 International Space Station (NASA Public Domain)

The International Space Station (ISS) provides a platform to conduct scientific research that cannot be performed in any other way. The primary fields of research include astrobiology, astronomy, human research including space medicine and life sciences, physical sciences, material science, space weather, and weather on Earth. Medical research improves knowledge about the effects of long-term space exposure on the human body, including muscle atrophy, bone loss, and fluid shift. The ISS provides a location in the relative safety of Low Earth Orbit to test spacecraft systems that will be required for long-duration missions to the Moon and Mars.

Scientific satellites perform space environment experiments in the fields of biology, chemical processes, metallurgy, botany, and medicine.

The Upper Atmosphere Research Satellite (UARS) shown in Figure 1-9 was a NASA-operated orbital observatory whose mission was to study the earth's atmosphere, particularly the protective ozone layer.

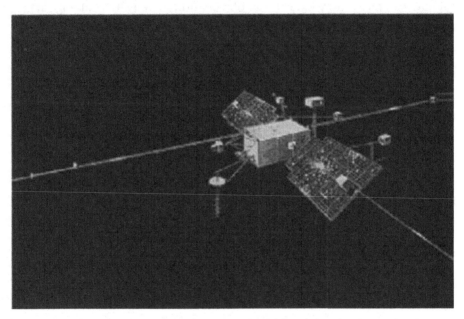

Figure 1-9 UARS Scientific Satellite (NASA Public Domain)

Measurements included: the concentrations and distributions of nitrogen and chlorine compounds, ozone, water vapor and methane; thermal emission from the **earth's limb** (edge of earth's disc as seen from space); hydrogen chloride, hydrogen fluoride, nitric oxide, nitrogen dioxide, temperature, aerosol extinction, aerosol composition, the emission and absorption lines of molecular oxygen above the limb of the earth, wind, and temperature and emission rate from airglow and aurora.

Satellites in Orbit

Objects in Orbit

Who keeps track of the objects orbiting the earth? The United States Space Surveillance Network (SSN) does. Its mission involves detecting, tracking, cataloging, and identifying artificial objects orbiting earth, including: active and inactive satellites, spent rocket bodies, or fragmentation debris.

The SSN tracked space objects since 1957 when the Soviet Union opened the space age with the launch of Sputnik I. Since then, the SSN tracked tens of thousands of space objects orbiting earth. Currently, the SSN tracks thousands of orbiting objects. The rest have re-entered earth's turbulent atmosphere and disintegrated, or survived re-entry and impacted the earth. The space objects now orbiting earth range from satellites weighing several tons to pieces of spent rocket bodies weighing only several pounds. A small percentage of the space objects are operational satellites, the rest are debris.

Numbers of Satellites

How many satellites are in orbit? As of October 1, 2011, there were 966 operating satellites in orbit. The United States, Russia, and China are the three countries with the most satellites owned outright. These three countries among them own 613 or about two-thirds of the active satellites. Who owns the other third? A number of other countries and partnerships own between 10-20 satellites, but at least 115 countries in total own a satellite or share in one.[2]

The four types of satellite orbits are: Low Earth Orbits (**LEO**) between 100 and 1,240 miles altitude, Medium Earth Orbits (**MEO**) between 1,243 and 22,236 miles altitude, **Elliptical** orbits' altitudes vary significantly between LEO and MEO altitudes during one orbit revolution, and Geostationary Orbits (**GEO**) at 22,236 miles altitude.

The total of 966 operating satellites is distributed into the following type of orbits:

- Low Earth Orbit (LEO): 470
- Medium Earth Orbit (MEO): 64
- Elliptical: 34
- Geostationary Orbits (GEO): 398

The total of 966 operating satellites is distributed by country as follows:

- United States: 443
- Russia: 101
- China: 69
- Others: 353

The total of 443 of U.S. satellites is distributed as follows:
- Commercial: 194
- Government: 128
- Military: 121

Out of the 398 total number of Geostationary Orbits (GEO) satellites, 355 are Communications Satellites which are distributed as follows:

- Commercial: 266
- Government: 30
- Military: 59

The 355 communications satellites in GEO orbit are distributed by country as follows:

- United States: 154
- Russia: 18
- China: 28
- Others: 155

Satellites in a geostationary orbit occupy a single orbit zone above the equator. Each satellite in geostationary orbit must be kept in its position within a sufficient distance so that it does not collide with other satellites.

Absent any other limitations on the availability of orbital positions, 1,800 satellites could be placed in a geostationary orbit at about 90 miles apart without posing a navigational hazard to others; however, as you shall see in the next section geostationary communications satellites group together over high population regions of the earth. To avoid harmful radio-frequency interference during operations the International Telecommunication Union assigns them an **orbital slot** (assigned longitude and radio frequency). This has led to conflicts among different countries wishing access to the same orbital slots and radio frequencies. These disputes are addressed through the International Telecommunication Union's allocation mechanism.[3]

Satellite Longitudes and Separation

How are satellites distributed and separated? Appendix B presents the number of geostationary (GEO) satellites in orbit and major cities that are within 15^0 longitude ranges. Table 1-1 presents a summary of Appendix B.

Table 1-1 GEO Satellites' Longitude Summary

Longitude Range	Number of Satellites	Region
180 W - 135 W	10	Alaska, Hawaii
135 W -60 W	102	Continental United States
60 W - 15 W	36	South America
15 W – 30 E	74	Europe, Africa
30 E - 90 E	83	Russia, Middle East, India
90 E - 180 E	75	Russia, China, Japan, Australia

Satellites group together above the high density regions such as above North America and Europe because of the high demand for communications in those regions.

Satellites that are close to each other have to maintain their longitude orbit positions very accurately to prevent a collision that could be caused by a thruster malfunction. Table 1-2 shows examples of **satellite clusters** that are closely grouped satellites in the same orbital slots.

Table 1-2 Satellite Clusters

Satellite	Operator	Longitude^0
Nimiq 2	Telsat Canada Ltd. - Canada	91.25 W
Nimiq 1		91.24 W
Nilesat101	Egyptian Radio and TV Union - Egypt	7.00 W
Nilesat201		7.00 W
Amos 3	Space Communication Ltd. - Israel	3.99 W
Amos 2		3.98 W
Eurobird9A	EUTALSAT (European Telecommunications Satellite Consortium) - Multinational	9.00 E
KA-SAT		9.00 E
Astra 1H	SES (Societe Europienne des Satellites) - Luxembourg	19.20 E
Astra 1M		19.20 E
Astra 1N	SES (Societe Europienne des Satellities) - Luxembourg	28.20 E
Astra 2C		28.21 E
Astra 2B		28.22 E
Arabsat 5A	ASCO (Arab Satellite Communication Organization) - Multinational	30.50 E
Arabsat 2B		30.51 E
Metsat-1	ISRO (Indian Space Research Organization) - India	74.07 E
INSAT 3C		74.08 E
GSAT-3		74.09 E
INSAT 4A	ISRO (Indian Space Research Organization) - India	83.08 E
INSAT 3B		83.09 E
BSAT-3B	Broadcasting Satellite Systems Corporation - Japan	110.00 E
BSAT-3C		110.00 E

Table 1-2 shows that eight satellites have no separation, but we know that there has to be some separation otherwise there would be a collision. The database[2] provided this information and research did not reveal the amount of separation. Given the distance around the geostationary orbit is 164,619 miles; there is 457 miles per degree of longitude.

Table 1-2 also shows that there are sixteen satellites that are only separated by 0.01^0 which equates to 4.57 miles. In addition, the database shows that there are 15, 8, 10 and 7 satellites that are separated by 0.20^0 (9.14 miles), 0.03^0 (13.72 miles), 0.04^0 (18.28 miles) and 0.05^0 (22.85 miles), respectively. Fortunately, the geostationary satellites are not exactly at the same altitude so they are also separated in altitude. What is the probability of a satellite collision in the geostationary orbit? Barring a control failure, collision probability for satellites in the geostationary belt is very small due to the fact that the satellites are moving in the same direction at a relatively low velocity (to each other).[4]

Debris in Orbit

How much debris is in orbit? Debris consists of everything from spent rocket stages and defunct satellites to erosion, explosion, and collision fragments. As the orbits of these objects often overlap the trajectories of newer objects, debris is a potential collision risk to operational satellites.

The vast majority of the estimated tens of millions of pieces of space debris are small particles, less than half an inch. Impacts of these particles cause erosive damage. Solar panels and optical devices (such as telescopes or star trackers) are subject to constant wear by debris and to a much lesser extent by micrometeorites.

Figure 1-10 shows the space debris populations seen from outside geostationary orbit (GEO). Note the two primary debris fields, the ring of objects in GEO and the cloud of objects in low earth orbit (LEO). In 2011, NASA said approximately 22,000 different objects were being tracked. More than 20,000 pieces are larger than a softball and it is estimated that 500,000 are larger than a marble. These objects travel around the earth at speeds up to 17,500 mph.[5]

ARTIFICIAL EARTH SATELLITES

The only protection against larger debris is to maneuver the satellite in order to avoid a collision. If a collision with larger debris does occur, many of the resulting fragments from the damaged satellite will be in the two pound mass range and these objects become an additional collision risk.

As the chance of collision is a function of the number of objects in space, there is a critical density where the creation of new debris occurs faster than the various natural forces that remove these objects from orbit. Beyond this point a runaway chain reaction can occur that reduces all objects in orbit to debris in a period of years or months.

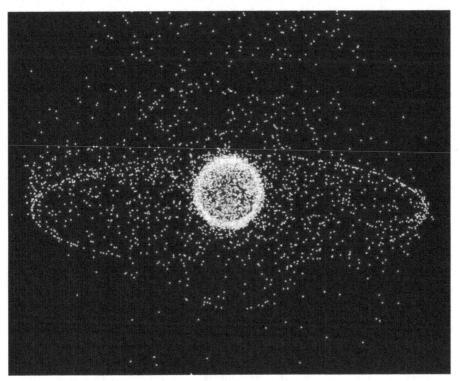

Figure 1-10 Space Debris (NASA Public Domain)

The 2009 satellite collision was the first accidental hypervelocity collision between two intact artificial satellites in earth orbit. The collision occurred on February 10, 2009 at 490 miles above the Taymyr Peninsula in Siberia, when Iridium 33 and Kosmos-2251 collided.

The satellites collided at a speed of approximately 26,170 mph, faster than escape velocity on earth. NASA reported that a large amount of debris was produced by the collision. As of March 2010, the U.S. Space Surveillance Network has cataloged 1,740 pieces of debris from the collision, with about 400 additional pieces waiting to be cataloged.

Events where two satellites approach within several miles of each other occur numerous times each day. Sorting through the large number of potential collisions to identify those that are high risk presents a challenge. Precise, up-to-date information regarding current satellite positions is difficult to obtain. Calculations had predicted these two satellites to miss by approximately 639 yards. An Iridium spokesman reported that they experienced close approaches, which numbered 400 per week (for approaches within 3 miles) for the entire Iridium constellation. He estimated the risk of collision per conjunction as one in 50 million.[6]

This collision and numerous near-misses have renewed calls for mandatory disposal of defunct satellites (typically by deorbiting them), but no such international law exists as of yet.

Satellite Operators/Owners

Who are the operator/owners of the satellites? Appendix C presents a list of the United States organizations that operate and/or own satellites. This list includes civil, commercial, government, military users and combinations thereof. In the civil arena we have the U. S. Air Force Academy, the U. S. Naval Academy, and several universities combined with government organizations.

In the commercial arena we have DirecTV, Globalstar, Iridium, Hughes Network Services, SAT-GE, SES, and Sirius Satellite Radio to name a few. Also, Intelsat operates a fleet of approximately fifty geostationary communications satellites.[7]

National Aeronautics and Space Administration (NASA) and National Oceanographic and Atmospheric Administration (NOAA) are the primary government operator/owners.

The military organizations include: Air Force Research Laboratory (AFRL), Defense Advanced Research Projects Agency (DARPA), Department of Defense (DoD), Strategic Space Command/Space Surveillance Network (SSN), US Air Force (USAF), and US Navy (USN).

Commercial communications satellite operators/owners determine market requirements for satellite services, identify customers' requirements for satellite communications in a geographic territory, finance the construction and launch of satellites, coordinate design and engineering with satellite manufacturers during satellite construction, place satellites into orbit through a launch provider, assume control of the satellite, and lease satellite capacity to customers for contracted periods.

Satellite capacity usually consists of one or more **transponders** (satellite communications channels), or a portion of one provided by the satellite operator. The operator controls the satellite through Tracking, Telemetry and Command (TT&C) earth stations and a satellite control center. The user chooses the satellite and the satellite operator to complete its network, and contracts with the satellite operator for the capacity.

Satellite Contractors

Who makes the satellites? Appendix C also presents a list of the contractors (domestic and foreign) for the U.S. Operator/Owners. This list includes several academic organizations such as the U. S. Air Force Academy, the U. S. Naval Academy, and several universities combined with government organizations.

In the commercial arena we have Ball Aerospace, Boeing, General Dynamics, General Electric, Hughes, Lockheed Martin, Loral Space Systems, SpaceQuest, Swales Aerospace, Northrop Grumman, Orbital Sciences Corporation, and TRW to name several. Some of these contractors also provide military satellites. National Laboratory contractors include: Los Alamos (LANL), Sandia (SNL) and Reconnaissance (RNL). NASA contractors include: Ames Research Center (ARC), Jet Propulsion Laboratory (JPL), and Goddard Space Flight Center (GSFC).

Satellite manufacturers are responsible for designing, fabricating, and testing the satellite according to the satellite operators'/owners' specifications and contracting for and procuring the needed components for building the satellite. They construct the satellite according to standards and specifications for withstanding the launch and space environments. They deliver the satellite to the launch provider and assist in the integration of the satellite with the launch vehicle.

Satellite Ownership Costs

How much do satellites cost? In the 2004 article "Satellite Costs"[8] David Green of Lockheed Martin Commercial Space Systems (LMCSS) presents the costs of satellite ownership. In addition to the purchase price of the satellite, costs include launch and on-orbit insurance, and operating costs. Premiums for a basic insurance package that includes launch, post-separation, commissioning, and first year of in-orbit operation of a commercial satellite is generally 18 percent to 22 percent of the purchase price based on satellite reliability. On-orbit insurance annual coverage costs an average of 2.5 percent to 3.0 percent of the purchase price. Fifteen years operating costs per satellite for mid-sized operators (3-12 satellites) are between $6M and $10M for staff and between $3M and $6M for infrastructure, resulting in a total operating cost of between $9M and $16M. By way of example, assuming a satellite purchase price of $100 million, the estimated total cost of ownership for a satellite for 15 years is as follows:

- Satellite Purchase Price $100M
- Launch Insurance $ 18M - $ 22M
- On- Orbit Insurance $ 35M - $ 42M
- Operating Cost $ 12M
- Total Ownership Cost $169M - $180M

The approximate ownership cost of a typical $250M geostationary communications satellite is $395M for a reliable satellite and $426M for a not so reliable satellite. This difference encourages satellite buyers to procure a reliable satellite. Launch costs are estimated in the next chapter.

2

Satellite Launch

Satellite Launch Vehicles

What boosts the satellites into space? Launch vehicles do. The primary satellite **launch vehicles** include: Atlas V, Delta IV, Proton, Ariane 5, and Long March. These launch vehicles come in several versions and can put satellite payloads in Low Earth Orbit (LEO), Medium Earth Orbit (MEO), Elliptical Orbits, and Geostationary Orbit (GEO). Multiple satellites have been launched into orbit by a single launch vehicle.

Primary satellite launch sites include:
- Kourou, French Guiana
- Cape Canaveral, Florida
- Vandenberg Air Force Base, California
- Baikonur, Kazakhstan
- Jiuquan Satellite Launch Center, China
- Taiyuan Satellite Launch Center, China
- Wenchang Satellite Launch Center, China
- Xichang Satellite Launch Center, China

Other satellite launch vehicles include the Minotaur IV, Pegasus, and Zenit. The Space Shuttle was decommissioned in 2011. Minotaur IV is an all-solid four stage vehicle that places US Government payloads into low earth orbits. Pegasus is a winged space launch vehicle, launched from an aircraft, capable of carrying small, unmanned payloads into low earth orbit. Zenit is a Russian launch vehicle that can be launched at sea or from land.

Atlas V shown in Figure 2-1 is an active expendable launch system in the Atlas rocket family. Atlas V was formerly operated by Lockheed Martin, and is now operated by the Lockheed Martin-Boeing joint venture United Launch Alliance (ULA).

Figure 2-1 Atlas V (Wikimedia Commons Public Domain)

Each Atlas V rocket uses a Russian-built engine to power its first stage and an American-built engine burning to power its Centaur upper stage. Some configurations also use strap-on booster rockets. In its more than two dozen launches, starting with its maiden launch in August 21, 2002, Atlas V has had a near-perfect success rate with only one anomaly.

Delta IV shown in Figure 2-2 is an active expendable launch system in the Delta rocket family. Delta IV uses rockets designed by Boeing's Integrated Defense Systems division that are built in the United Launch Alliance (ULA) facility in Decatur, Alabama.

Figure 2-2 Delta IV (Wikimedia Commons Public Domain)

ULA completes final assembly at the launch site. The rockets launch payloads into orbit for the United States Air Force and commercial satellite users. Delta IV rockets are available in five versions which are tailored to suit specific payload size and weight ranges. Delta IV was primarily designed to satisfy the needs of the U.S. military.

Proton shown in Figure 2-3 is an expendable launch system used for both commercial and Russian government space launches. The first Proton rocket was launched in 1965 and the launch system is still in use as of 2012, which makes it one of the most successful heavy boosters in the history of spaceflight.

Figure 2-3 Proton (Wikimedia Commons Public Domain)

All Protons are built at the Khrunichev plant in Moscow, and then transported for launch to the Baikonur Cosmodrome, where they are brought to the launch pad horizontally and then raised into vertical position for launch. Commercial launches are marketed by International Launch Services (ILS).

Ariane 5 shown in Figure 2-4 is a series of European civilian expendable launch vehicles for space launch use. Arianespace launches Ariane rockets from the Centre Spatial Guyanais at Kourou in French Guiana where the proximity to the equator gives a significant advantage for the launch.

Figure 2-4 Ariane 5 (Wikimedia Commons Public Domain)

There are five versions of Ariane. Ariane 1 was a three-stage launcher, derived from missile technology. Arianes 2 through 4 are enhancements of the basic vehicle. The largest versions can launch two satellites. Ariane 5 is a nearly-complete redesign. Two solid-fuel boosters are strapped to the sides.

Long March shown in Figure 2-5 is China's primary expendable launch system family. The Long March family has 7 members. Long March 2 through 4 are active and Long March 5 is under development. Long March 6 and 7 are derivations of Long March 5.

Figure 2-5 Long March 3 (Wikimedia Commons Public Domain)

Long March 5 is a Chinese next-generation heavy lift launch system that is currently under development by China Academy of Launch Vehicle Technology (CALT). China markets launch services under the China Great Wall Industry Corporation. There are four launch centers located in China.

Table 2-1 presents launch vehicles' number of stages, launch mass, and maximum satellite payload mass capabilities for Low Earth Orbit (LEO) at 100 – 1,240 miles altitude and Geostationary Orbit (GEO) at about 22,236 miles altitude. To prevent confusion, it is important to understand the two types of payloads. A **satellite payload** is the satellite; whereas, the **mission payload** is the onboard equipment for performing the satellite mission.

Table 2-1 Launch Vehicles Capabilities

| Launch Vehicle | Stages | Total Mass | Max. Satellite Payload Mass | | Mass Ratio* | |
		Launch pounds	LEO pounds	GEO pounds	LEO	GEO
Atlas V	2	737,400	64,820	28,660	11	26
Delta IV	2	530,000- 1,616,800	49,470	28,620	33	56
Proton	3	1,529,600	101,388	24,245	15	63
Ariane 5	2	1,712,000	46,286	23,143	37	74
Long March	2 or 3	423,287- 760,593	26,449	12,123	29	63
*Mass Ratio is calculated as the Maximum Launch Mass divided by the Maximum Payload Mass.						

The difference between the total launch mass and satellite payload mass is mostly propellant. The **mass ratio** gives an approximate measure of how many pounds of propellant it takes to launch one pound of satellite. Table 2-1 shows that out of the four launch vehicles that can launch over 20,000 pounds into geostationary orbit, the Atlas V launch vehicle has the lowest mass ratio. The average mass ratio of these four launch vehicles is approximately 55.

Appendix D presents a list of launch vehicles, with their launch sites, used by U.S. Operator/Owners. Atlas and Delta launch vehicles are launched from Cape Canaveral in Florida and Vandenburg Air Force Base in California. Ariane is launched from the Guiana Space Center in Kourou, French Guiana. Proton is launched from the Baikonur Cosmodrome in Baikonur, Kazakhstan. And Long March is launched from the Xichang Satellite Launch Center, China.

Other Launch Systems

Minotaur IV is an all-solid four stage vehicle. It uses the first three stages of the Peacekeeper missile, topped with an additional fourth stage, to place satellites into low Earth orbits.

Orbital Sciences Corporation developed and operates the Minotaur IV. Because it contains missile components which were provided by the U.S. Government, the Minotaur IV cannot be used for commercial launches and hence it is marketed towards U.S. Government payloads. Minotaur IV launch sites include: Wallops Island Flight Facility, Kodiak Launch Complex, and Vandenberg Air Force Base.

Pegasus is a winged space launch vehicle capable of carrying small, unmanned payloads of 980 pounds into low earth orbit. It is air-launched as part of an expendable launch system developed by Orbital Sciences Corporation.

Three main stages burning solid propellant provide the thrust. It flies as a rocket-powered aircraft before leaving the atmosphere. A carrier aircraft carries the Pegasus to 40,000 feet for launch. The carrier aircraft provides flexibility to launch the rocket from anywhere rather than just a fixed pad.

Zenit (Zenith in English) is a Russia launch vehicle that can be launched at sea or from land. Its launch mass is over a million pounds.

Payload capability is about 30,000 pounds to low earth orbit and about 11,000 pounds into geostationary orbit. The Zenit-3SL is launched by the Sea Launch Consortium's floating launch platform in the Pacific Ocean.

Sea Launch rocket assembly occurs in Long Beach, California. Launches occur from the *Ocean Odyssey* offshore launch platform, situated at the equator. *Ocean Odyssey* is also used to transport rockets to the launch site. Zenit-2 is launched from Baikonur Cosmodrome in Kazakhstan.

Vehicle Configurations

Figure 2-6 shows the configurations a geostationary communications satellite goes through from launch to operational orbit.

The satellite is in a **stowed** configuration with folded solar arrays and reflector antennas for launch and during transfer to its operational orbit. During launch the satellite fits within a protective covering called a **launch shroud.**

The first two stages boost the third stage and payload to a sub orbital altitude. After jettisoning the shroud, the third stage thrusts the payload into a **transfer orbit,** which is an orbit that moves a satellite from one orbit to another.

After this maneuver the satellite separates from the third stage and is then on a transfer orbit to the geostationary orbit altitude.

After reaching geostationary altitude it performs maneuvers to get into the geostationary orbit, deploys the antennas and solar arrays, establishes the correct orientation, establishes communication links, and becomes fully operational.

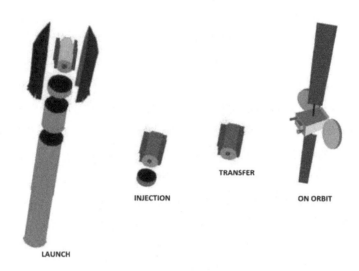

TRANSFER

INJECTION

ON ORBIT

LAUNCH

Figure 2-6 Satellite Configurations

Launch Costs

How much does it cost to launch a satellite? The estimated launch costs[9] for the 5 major launch vehicles are as follows:

- Atlas V $85M - $135M
- Delta IV $70M - $140M
- Proton $80M - $105M
- Ariane 5 $120M
- Long March 3 $35M - $ 40M

The launch costs vary due to different launch vehicle configurations for different satellite payloads. Some of these vehicles can launch multiple satellites allowing the launch costs to be prorated according to satellite mass. For example, if Ariane 5 launched two satellites each weighing 11,000 pounds, they would each pay half of the launch costs or $60M. Adding $60M to the satellite ownership cost of a reliable geostationary communications satellite from the previous chapter of $399M, results in a mission cost of $459M.

3

Geostationary Communications Satellite

The remainder of this book focuses on the satellite that affects our daily lives and our security---the Geostationary Communications Satellite.

As stated above, this satellite provides us with global television, radio, business, Internet, and telephone services. Communications satellites are critical for our national security because they provide global command and control of our military resources.

We need to protect these satellites from attack by **killer satellites** and missiles launched from the ground. We have backup **terrestrial** (earth) communication systems, but our communication capabilities would be significantly diminished if we lost our communications satellites.

Because radio signals cannot curve along the curvature of the earth we use satellites for communications. We send radio signals up to the satellite in a straight line and receive them down from the satellite in a straight line over a large coverage area. This circumvents the need for a large network of terrestrial **signal relay stations** and increases range.

Geostationary communications satellites receive radio frequency signals from the earth, amplify them, shift their frequencies, and transmit them back to the earth. They are placed in a geostationary orbit at 22,236 miles above the equator at specific longitudes or **stations** such that their rotation around the earth is synchronous with the rotation of the earth.

They remain on this station for their operational lifetime, receiving and transmitting signals from and to all communication points in their earth coverage area.

Because the distance the radio signals travel to and from the satellite is greater than the distance they travel on the surface of the earth, there is a delay or **latency** in when they reach the users.

Satellite **perturbations** include: gravitational attractions of the moon, the sun, solar winds, meteorites, and **earth anomalies** or the non-uniform mass distribution of the earth. We design communications satellites to perform their communication functions, point in the right direction, maintain **attitude** or orientation, and stay on station over the equator in the face of these perturbations.

Booster rocket engines launch the satellites into space and transfer orbits get them to the geostationary orbit. The required **speed maneuvers** for these orbits determine the amount of propellant fuel needed. Command and control of the satellite comes from **earth stations**.

Satellite lifetime in orbit is dependent on component redundancy and available propellant fuel. Component redundancy gives required reliability for the target satellite lifetime. The life of the satellite is directly related to the propellant fuel available for **attitude control** or pointing the satellite in the right direction and compensating for rotational perturbations and **station-keeping** or keeping the satellite in the right orbit so that it stays on its assigned longitude above the equator.

Original communications satellite configurations were dual spin stabilized. In this configuration, an outer cylinder covered with solar cells spins to provide gyroscopic attitude stability to the satellite. The inner platform with mounted antennas keeps its attitude pointing at the center of the earth. A dual spin joint transfers the electrical current generated by the spinning solar array to the satellite stable platform. Dual-spin satellites gave way to **three-axis stabilized** satellites starting in the 1970's due to the overall more cost-effective performance of the three-axis stabilized satellites.

Some companies have their own fleet of satellites. Other users lease part of the communications capabilities of a satellite. Some satellites cover a domestic region, while others cover more than one country. Some large television broadcast companies need high reliability of service delivery and long lifetime. They consider redundancy at all levels, including: on-board component redundancy, contingent lease agreements with other providers, and launch-ready backup satellites.

Trade studies, contingency plans and risk analyses raise the confidence in satisfying the companies' needs. Of course these companies are also interested in obtaining the most cost-effective solution to their business requirements. So they look for solutions that provide maximum revenue, uninterrupted service, and long lifetime with lowest implementation, operation, and maintenance costs.

Geostationary Orbit

What is a geostationary orbit? Geostationary communications satellites are in a circular geostationary orbit which is in the same plane as the equator. As shown in Figure 3-1 the satellite rotates around the earth at the same rate as the earth rotates around its axis and stays on station above the same point at the equator. The geostationary orbit plane is aligned with the **equatorial plane** making the **inclination angle** between the two planes zero degrees.

In this orbit the satellite points its antennas at the earth to establish a beam that provides a coverage area or **footprint** on the earth's surface. The altitude of the satellite in the circular geostationary orbit is about 22,236 miles above the earth and it travels in this orbit at about 6,878 miles per hour. This altitude is about 5.6 times the radius of the earth and the speed is about 105 times our legal freeway speed.

At geostationary altitude the earth appears as a disc of about 17.3^0 wide. Typical **antenna beams** are about 14^0 or less. The satellite's attitude is controlled to keep the beam pointing within 0.03^0 to 0.05^0. The solar arrays rotate to face the sun for maximum electrical power generation.

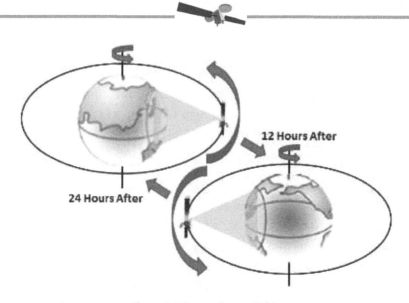

Figure 3-1 Geostationary Orbit

Satellite Communications Services

What services are provided by communications satellites? Figure 3-2 shows a geostationary communications satellite providing the above mentioned television, radio, business, Internet, and telephone services on the land, in the air, and on the sea.

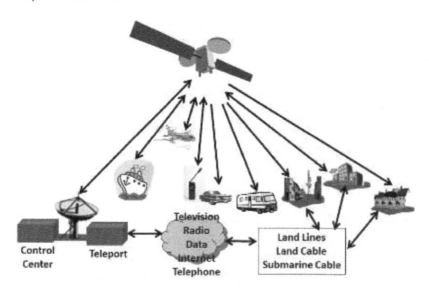

Figure 3-2 Satellite Communications Services

Satellite Television

North American television and radio uses two types of satellite television service: Direct Broadcast Satellite (DBS) and Fixed Service Satellite (FSS).

A direct broadcast satellite is a communications satellite that transmits to small DBS satellite dish antennas (usually 18 to 24 inches) commonly called Very Small Aperture Terminal (VSAT) antennas. DBS technology is used for Direct-To-Home (DTH) satellite TV services, such as DirecTV and DISH Network in the United States, Bell TV and Shaw Direct in Canada, Freesat and Sky Digital in the UK, the Republic of Ireland and New Zealand, and DSTV in South Africa.

Fixed Service Satellites normally provide: **broadcast feeds** to and from television networks and local affiliate stations (such as program feeds for network and syndicated programming); live shots and **backhauls**, which get audio and video material to a main distribution point; and for distance learning by schools and universities.

FSS satellites usually distribute free-to-air satellite TV channels. FSS satellites require a much larger reception dish (3 to 12 feet or larger) than the DBS receiving antennas.

Mobile Television

Special antennas exist for mobile reception of DBS television. Using Global Positioning System (GPS) technology as a reference, these antennas automatically aim to the satellite no matter where or how the vehicle (on which the antenna is mounted) is situated. Ships and recreational vehicles use these mobile satellite antennas. JetBlue Airways for DirecTV (supplied by LiveTV, a subsidiary of JetBlue) also uses mobile DBS antennas.

Satellite Radio

Satellite radio offers audio services in some countries, notably the United States.

A satellite radio or Subscription Radio (SR) is a digital radio signal that is broadcast by a communications satellite, which covers a much wider geographical range than terrestrial radio signals. Mobile services, such as Sirius, XM and Worldspace, allow listeners to roam across an entire continent, listening to the same audio programming anywhere they go.

Satellite Business

Businesses and government agencies private networks communications services include: corporate wide-area networking, in-store digital signage, inter-office and remote site networking for branches and teleworkers, Point of Sale (POS) transactions, back-office store services, distance learning applications, and video conferencing. Service to sites that do not have access to terrestrial forms of connectivity, include: construction sites, drilling platforms, and temporary facilities in remote places.

Satellite Internet

Compared to ground-based communication, all geostationary satellite communications experience high latency due to the signal having to travel more than 22,236 miles to a satellite in geostationary orbit and back to earth again. Even at the speed of light of 186,000 miles per second, this delay is about 0.25 seconds.

With old less efficient modems round trip latency can be as much as 1.2 seconds or more. New more efficient modems add less latency to the link. Round trip latency is typically about 0.65 seconds. This amount of latency makes satellite Internet unattractive for some recreational activities like real time interactive video gaming.

Satellite Internet service users include businesses and remote area users where telephone, cable, or fiber optic lines are not available. Ground telephone line and cable broadband Internet users usually experience about 0.03 to 0.30 seconds latency. In addition to the lower latency, ground Internet service is less expensive than satellite Internet service.

Satellite Telephone

Land lines and submarine communications cables carry most of the world's telephone calls.

However, certain areas use satellite fixed telephony, including: regions of some continents and countries where landline telecommunications are rare to nonexistent; for example large regions of South America, Africa, Canada, China, Russia, Australia, and the edges of Antarctica and Greenland. Also, satellite telephone serves remote islands such as Ascension Island, Saint Helena, Diego Garcia, and Easter Island, where no submarine cables are in service.

The fixed Public Switched Telephone Network (PSTN) relays telephone calls from land line telephones to an earth station, where they are then transmitted to a geostationary satellite. The downlink follows an analogous path. Satellite phones connect directly to a constellation of either geostationary or low earth orbit satellites. Calls are then forwarded to a satellite telecommunications port (teleport) connected to the PSTN.

Satellite Disaster Recovery

Satellites provide communications services for disaster recovery when catastrophe strikes cutting off emergency terrestrial communications. Satellites can provide internet access, Voice over Internet Protocol (VoIP), and other network services when other wired and wireless services are impaired by natural disaster.

Satellite Internet Protocol Television

Satellites provide Internet Protocol Television (IPTV), which is gaining popularity among businesses and consumers by providing the potential for new viewing experiences, interactivity, and access to a virtually unlimited amount of content.

Ground Stations

Telecommunications Ports (Teleport)

Telecommunications Ports or Teleports consist of several large antennas and buildings containing transmitters and receivers and signal processing, routing, and interface equipment. Globally distributed Teleports provide customer communications services by accessing satellites. They process and route the signals they transmit and receive from the satellite and from terrestrial communications segments carried on land lines, land cable, and submarine cable.

Tracking, Telemetry and Command

The ground Tracking, Telemetry and Command (TT&C) system tracks the satellite's orbital position and receives satellite status information via telemetry pertaining to payload health, satellite attitude, power production and storage, propellant remaining, equipment failures, thermal conditions, and performance anomalies. Based on this information the TT&C sends commands to the satellite to control the satellite and correct any problems or anomalies. The TT&C also commands and controls the satellite orbit change and attitude pointing maneuvers.

Satellite Components

What are the components of a communications satellite? Figure 3-3 shows typical communications satellite components. The mission payload consists of the communications equipment including the antennas and **repeater** containing the transponders.

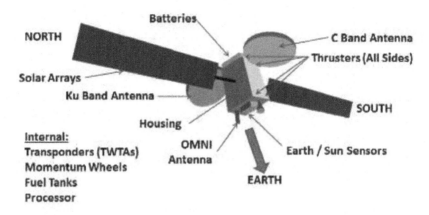

Figure 3-3 Satellite Components

The rest of the satellite is called the satellite **bus**. The bus transports the communications payload around in its orbit, provides electrical power, maintains attitude, points the payload, keeps the satellite on station, and makes orbit changes. The thermal control subsystem keeps the electronics and other components within a safe temperature range over the life of the satellite. The structure holds everything together and protects the components during launch and after **deployment** on orbit.

An onboard beacon transmits signals to a ground Tracking, Telemetry, and Command (TT&C) Station. The onboard **telemetry** system reports health and status of all onboard systems and receives commands to control all onboard systems in support of the payload mission to provide uninterrupted communication services for all users. An **Omni** antenna provides communications connectivity during all phases of the mission.

Typical satellite on-orbit lifetime is ten to fifteen years which is possible with redundancy. Usually the limiting factor of lifetime is the amount of propellant fuel remaining. A significant number of communications satellites in orbit are functional, but are inoperable because they ran out of fuel.

Communications

Most of us use electrical alternating current every day. Electric current in the wire flows in one direction for a period of time and then in the opposite direction for the same period of time (called a cycle) to produce the alternating current. Electrical current oscillates at 50 or 60 cycles per second (Hz). The **hertz** (symbol **Hz**) is the international unit of frequency defined as the number of **cycles per second** of a periodic phenomenon. Commonly used multiples of hertz are kHz (kilohertz, thousand Hz), MHz (megahertz, million Hz), GHz (gigahertz, billion Hz) and THz (terahertz, trillion Hz).

Radio frequency (**RF**) is a rate of oscillation in the range of about 3 kHz to 300 GHz, which corresponds to the frequency of radio waves, and the alternating currents which carry radio signals. Electric currents that oscillate at radio frequencies have special properties not shared by alternating current at lower frequencies. The energy in an RF current can radiate off a conductor (antenna) into space as electromagnetic waves (radio waves); this is the basis of radio technology. RF current does not penetrate deeply into electrical conductors but flows in a thinner layer of the conductor, closer to the surface; this is known as the **skin effect**.

Radio waves carry information by varying a combination of: amplitude, frequency, and phase of the wave within a **frequency band**. In satellite communications, the transmission of most radio waves between the ground and satellite is at frequencies in the C and Ku frequency bands as listed in Table 3-1.

Table 3-1 Communication Frequency Bands

Frequency Band	Frequency Range (GHz)
C	4 - 8
Ku	12-18 (12.5 – 18 in North America)

Ground Antennas

Ground antennas used for receiving satellite signals and transmitting to the satellites vary considerably according to their application. Parabolic reflectors are the most widely used. The sizes of the antennas vary considerably. The parabolic reflectors used for satellite television reception are very small. However antennas used for professional applications, like transmitting programs up to the satellite where a much higher signal level is required to ensure the best possible picture is radiated back to earth, are much larger and may range up to several tens of feet in size.

Do you remember seeing large antennas (3 to 12 feet) in people's yards? That's because originally C-band was used for direct broadcast television. Going to the higher frequency Ku-band resulted in the smaller antennas (18 to 24 inches) we now have on our roofs, as shown in Figure 3-3, for receiving direct broadcast television.

Figure 3-4 Ku-Band Antenna

Compared with C-band, Ku-band is not restricted in power to avoid interference with terrestrial microwave systems, thus allowing the power of its uplinks and downlinks to be increased.

This higher power translates into smaller receiving dishes and points out a generalization between a satellite's transmission power and a dish's size. As the power increases, the dish's size can decrease.

This is because the purpose of the dish element of the antenna is to collect the incident waves over an area and focus them all onto the antenna's actual receiving element, mounted in front of the dish and pointed back towards its face. If the waves are more intense, fewer of them need to be collected to achieve the same intensity at the receiving element.

The satellite operator's earth station antenna does require more accurate position control when operating at Ku-band than when operating at C-band. Position feedback accuracies are higher and the antenna may require a closed loop control system to maintain position under wind forces acting on the dish surface.

There are, however, some disadvantages of Ku-band systems. Especially at frequencies higher than 10 GHz in heavy rain fall areas, a noticeable degradation occurs, due to the problems caused by and proportional to the amount of rainfall (commonly known as **rain fade**). Applying an appropriate link budget strategy when designing the satellite network and allocating higher power consumption compensates for rain fade loss.

A similar phenomenon, called **snow fade** (where snow or ice accumulation significantly alters the focal point of a dish) can also occur during winter precipitation. Under both rain fade and snow fade conditions, Ku-band losses can be reduced using effective coatings.

In satellite telecommunications terminology, **uplink** means the signal sent from earth to the satellite and **downlink** means the signal from the satellite to earth. The uplink and downlink signals to and from a specific satellite must be different. The satellite constantly sends a specific downlink frequency that identifies it from other satellites. The transmit signal must be strong enough to reach the earth stations. An earth station constantly sends a specific uplink frequency that identifies it from other stations.

The transmit signal must be strong enough to reach the satellite. The received signals at the earth station and the satellite are very weak compared to the transmitted signals. If uplink and downlink frequencies were the same, the transmitted signals would drown out the signals from the opposite ends.

By the time the uplink signal reaches the satellite it is weak and must be amplified before it is retransmitted to the earth.

Table 3-2 presents typical uplink and downlink bands for the C and Ku frequency bands.

Table 3-2 Downlink and Uplink Bands

Frequency Band	Downlink Band (GHz)	Uplink Band (GHz)
C	3.7 – 4.2	5.925 – 6.425
Ku	10.9 – 12.75	14 – 14.5

Mission Payload

The communications mission payload consists of the antennas and the repeater. The antennas on the satellite create footprint coverage and the repeater receives and transmits the signals from and to the ground.

Antennas on the Satellite

The distances that signals travel to geostationary satellites are very large. This means that path losses are high and signal levels are low. In addition to this the power levels that can be transmitted by satellites are limited by the fact that all the power has be generated from solar panels.

As a result the antennas are usually high gain directional parabolic reflectors. We want the satellite antennas to be as large as possible to receive low level signals and to transmit high level signals; however, the antennas on board the satellite are typically limited in size to around 7 to 10 feet by the space that is available on the satellite structure. Most communications satellites have either C-band or Ku-band antennas or both. Antennas consist of reflectors, feeds, feed networks, support structure, and pointing mechanisms.

Downlink Footprint

A **footprint** is the geographical area that can be served by a satellite. An example footprint is shown in Figure 3-5.

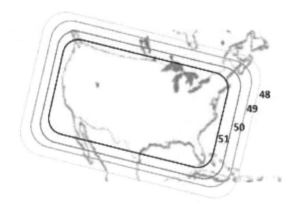

Figure 3-5 Antenna Footprint

The most important technical characteristic of a communications satellite is the transmit power for the downlink into the coverage area, which determines the overall quality of the link between the satellite and the earth station in addition to the size (and cost) of receiving dishes. The higher the Effective Isotropic Radiated Power (**EIRP**) in the direction of the receiving dish means a better quality link, or smaller receiving dishes. Measurement of EIRP is in **decibels** relative to one watt, or **dBW**. One watt is equivalent to 0 dBW and 50 dBW is equivalent to 100,000 watts.

Proper design of the satellite's coverage pattern produces maximum power or EIRP over a country, continent, region or zone without wasting energy over the water or land outside the footprint. Table 3-3 presents the minimum size of the ground Ku-band antennas for the EIRP values within the contour areas shown in Figure 3-5.

Table 3-3 Antenna Size for EIRP

EIRP	Antenna Diameter
51 dBW	12.5 inches
50 dBW	14.5 inches
48 – 49 dBW	18.0 inches

Uplink footprint

The uplink provides the satellite with signals to retransmit into the downlink. Like the downlink, the uplink has a coverage footprint. The receive footprint is specified with two parameters - the gain to noise temperature ratio (G/T), which is also called the receive figure of merit, and the Saturation Flux Density (SFD).

Without trying to explain these in detail, they are used to determine the size of the uplink antenna and transmit power. In a network of two-way Very Small Aperture Terminals (VSATs), the SFD determines the required VSAT transmit power, which determines the type, and therefore the cost, of the required solid state power amplifier on the VSAT antenna.

Transponders

A satellite's communications channels are called transponders. Most communications satellites are microwave radio relay stations in orbit and carry dozens of transponders, each with a bandwidth of tens of megahertz. Most transponders operate on a **bent pipe** principle, referring to the sending back of what goes into the conduit with only amplification and a shift from uplink to downlink frequency.

With data compression and multiplexing, several video and audio channels may travel through a single transponder on a single wideband carrier. Simply speaking, transponders provide the signal amplification and frequency shift.

As shown in Figure 3-6 each transponder has an input filter to separate out the particular bandwidth that is allocated to it, a preamplifier, and a RF **Traveling Wave Tube Amplifier (TWTA)**, and an output filter.

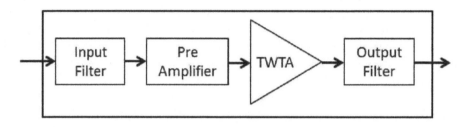

Figure 3-6 Transponder

Repeaters

The repeater contains microwave receivers, RF multiplexers, power amplifiers, channel processing and switching circuits, and transponders. Figure 3-7 shows how a group of transponders share the wideband components (the antenna and wideband receiver) to form a repeater. Each transponder and its components have a unique number. A single transponder can relay as much as 90 **Mbps** (megabits per second) with a suitable antenna, power amplifier, and modulation equipment on the ground.

Normally, a transponder carries several Very Small Aperture Terminal (VSAT) networks or up to 12 digital TV channels. Some Ku-band satellites carry 24 to 52 transponders. The repeater includes a quantity of spare TWTAs to cover amplifier failures.

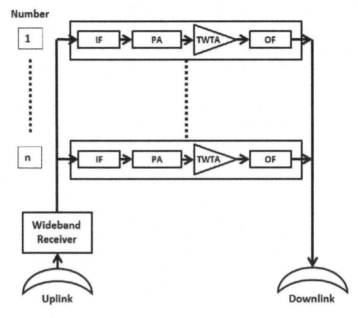

Figure 3-7 Repeater

Traveling Wave Tube

A **Traveling-Wave Tube (TWT)** is an electronic device that amplifies radio frequency signals to high power. The bandwidth of a broadband TWT can be as high as one octave, with operating frequencies in the 300 MHz to 50 GHz range. The voltage gain of the tube can be of the order of 70 decibels (10 million times).

The TWT shown in Figure 3-8 is an elongated vacuum tube with an electron gun (a heated cathode that emits electrons) at one end. A magnetic containment field around the tube focuses the electrons into a beam, which then passes down the middle of a wire helix that stretches from the radio frequency (RF) input to the RF output. The electron beam finally strikes a collector at the other end.

A directional coupler, which can be either a waveguide or an electromagnetic coil fed with the low-powered radio signal that is to be amplified, positioned near the emitter, induces a current into the helix. The helix acts as a delay line, in which the RF signal travels at nearly the same speed along the tube as the electron beam.

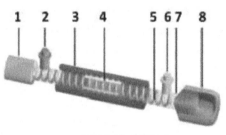

Cutaway View

(1) Electron Gun   (2) RF Input

(3) Magnets   (4) Attenuator

(5) Helix Coil   (6) RF Output

(7) Vacuum Tube   (8) Collector

Figure 3-8 Traveling Wave Tube (Wikimedia Commons Public Domain)

The electromagnetic field due to the RF signal in the helix interacts with the electron beam, causing bunching of the electrons (an effect called velocity modulation) and the electromagnetic field due to the beam current then induces more current back into the helix.

A second directional coupler positioned near the collector, receives an amplified version of the input signal from the far end of the helix. An attenuator placed on the helix, usually between the input and output helixes, prevents reflected waves from traveling back to the cathode.

Satellite Bus

Electrical Power

How do you generate electrical power on a satellite? Solar cells arranged in solar modules generate electrical power. Several solar modules form a solar panel. Several solar panels make up two solar arrays attached to opposite sides of the satellite. Batteries provide power before the solar panels are deployed, when the satellite is in the earth's shadow, and under peak load demands.

A **solar cell** (also called photovoltaic cell or photoelectric cell) is a solid state electrical device that converts the energy of light directly into electricity by the photovoltaic effect.

The solar cell works in three steps:

1. Photons in sunlight hit the solar panel and are absorbed by semiconducting materials, such as silicon.

2. Electrons (negatively charged) are knocked loose from their atoms, allowing them to flow through the material to produce electricity. Due to the special composition of solar cells, the electrons are only allowed to move in a single direction.

3. An array of solar cells converts solar energy into a usable amount of direct current (DC) electricity.

Solar cells are electrically connected and encapsulated as a **solar module**. Solar modules often have a sheet of glass on the front (sun up) side, allowing light to pass through while protecting the semiconductor wafers from abrasion and impact due to space debris and meteorites. Solar cells are also usually connected in series in modules, creating an additive voltage. Connecting cells in parallel will yield a higher current. Modules are then interconnected, in series or parallel, or both, to create a **solar panel** with the desired peak direct current (DC) voltage and current. Several solar panels combine to make up two **solar arrays**.

As a satellite travels along the orbital path, the solar arrays must remain pointed at the sun to produce power. This is accomplished by a number of methods based on the satellite design. Satellites with fixed arrays must point their surfaces at the sun or spin around the axis that allows them to rotate into the sunlight. With movable solar arrays, they either actively track the sun or are driven at a constant rate to maintain their pointing.

Solar arrays on a three axis stabilized satellite, where stabilization is achieved by controlling the rotation of the satellite about all three axes, a positioning mechanism that either steps (incrementally rotates) the arrays or actively positions them based on keeping peak power output.

The angular step size of a stepper motor, the gear ratio, and the angular change needed determines the number of steps, which determines the required number of pulses per second. The solar array drive electronics generates the pulses and applies them to the motor to complete a 360 degree revolution a day.

Active positioning systems also employ the use of stepper motors by converting solar array samples to pulses and applying them to the stepper motor to maintain the peak output power.

The output of a sun sensor on the solar array gives the direction to the sun for pointing the solar arrays. During eclipse periods the system normally reverts to a stepped mode and the active control resumes when the satellite returns to full sunlight.

Batteries provide power for the time when the satellite is in the earth's shadow. The communications equipment uses most of the power (approximately three-quarters). The amount of power required on board will vary considerably from a few hundred to several thousand watts depending on the mission payload.

The attitude of the satellite is controlled so that the communication antennas point at the earth and the axis of the solar arrays are aligned in the north-south direction perpendicular to the orbit plane.

A **sun sensor** provides the direction to the sun. Figure 3-9 shows the solar arrays rotate around their axis to always point at the sun.

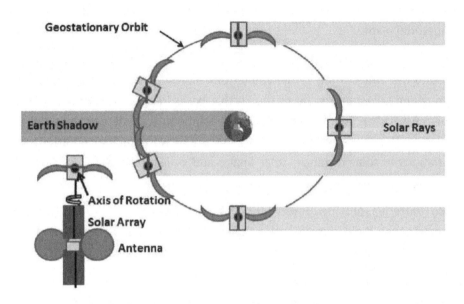

Figure 3-9 Sun Pointing Solar Arrays

The solar arrays on geostationary satellites are subject to a number of factors which can result in significant fluctuations in the amount of power available to onboard systems.

To begin with, the position of the satellite relative to the sun varies throughout the year because the earth's distance to the sun varies. Also, the plane of the earth's equator does not lie in the plane of the earth's orbit around the sun known as the **ecliptic** plane.

The sun moves from 23° below the equatorial plane at the winter **solstice** to 23° above the equatorial plane at the summer solstice and back again over the course of a year.

This motion changes the angle of incidence of solar energy received on the solar arrays since they must rotate about an axis perpendicular to the equatorial plane.

The geostationary orbit is outside the cone of the earth's shadow until around the times of the vernal and autumnal **equinoxes** at the beginning of spring and fall when the tilt of the Earth's axis is inclined neither away from nor towards the Sun. At these times, geostationary satellites can spend as much as 70 minutes of every day in shadow. If we combine the effects of variations in solar distance, solar angle, and eclipses over the course of a year, the total solar energy available varies by 12%.

Two batteries provide power during launch, ascent, and orbit transfer phases of the flight prior to antenna and solar array deployment, when the satellite is in the earth's shadow, and under peak load demands.

Automatic capabilities provide eclipse-load disconnect/reconnect control and battery under-voltage disconnect. Both features can be overridden by manual command. The satellite load condition in sunlight is such that some equipment must be powered off during eclipse. Ground command can enable or override automatic load shedding.

An automatic load shedding capability can be enabled and/or overridden upon ground command. Satellite telemetry monitors individual cell voltages in each battery. Periods of cold exposure require battery heaters. Onboard thermostats control battery temperatures and satellite telemetry monitors temperatures. Manual commands can override onboard battery heater thermostats.

Attitude Control

How do you point the satellite in the right direction to perform the satellite's mission? Figure 3-10 shows the satellite's **reference axes** of rotation around its center of mass. The **yaw axis** points to the center of the earth. The **roll axis** points along the direction of travel. The **pitch axis** is orthogonal or at a right angle to the roll-yaw (equatorial) plane. The roll, pitch, and yaw axis reference frame was originated for ships and then applied to aircraft and satellites.

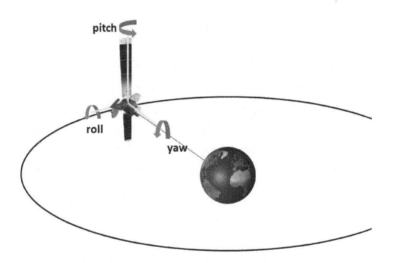

Figure 3-10 Satellite Axes of Rotation

Another way to visualize the reference axes is to imagine that you are on a cruise ship that is sailing on the edge of an orbit as shown in Figure 3-11 on the next page. You are looking down at the North Pole. The roll axis is in the direction the ship is sailing, so when the ship rolls it is about this axis. The pitch axis comes out of the page, so when the bow of the ship pitches up and down it is about this axis. Likewise the ship yaws from side to side when it rotates about the yaw axis.

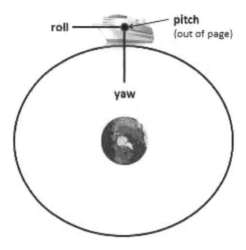

Figure 3-11 Ship Sailing on Edge of Orbit

Attitude control is the purposeful manipulation of controllable external forces to establish a desired attitude, whereas attitude determination is the utilization of vehicle sensors to ascertain the current vehicle attitude. The satellite's attitude control system points the satellite reference axes in specified directions, as a function of time and with specified accuracy, to satisfy mission objectives.

Geostationary communications satellites must point at the earth with high accuracy to keep an antenna beam coverage pattern over the specified area of the earth's surface. Also, the axis of the solar arrays must be maintained in a north-south orthogonal or right angle direction to the equatorial plane to generate power for the payload and the rest of the satellite or bus. Gravitational forces of the sun and moon, solar wind, meteorites, and earth anomalies perturb or disturb this nominal satellite attitude. The attitude control system corrects these perturbation torques.

The attitude control system consists of equipment to measure and maintain or change the orientation of the satellite. The earth and sun sensors measure and report the deviations of the satellite's required orientation to an onboard Controller that issues command signals to the **momentum wheels** to correct for any external torque perturbations to the satellite.

Satellites use a variety of attitude sensors based on mission requirements, including: gyroscopes, star sensors, **earth sensors**, and sun sensors. Geostationary communications satellites typically use earth and sun sensors mounted along the yaw axis as shown in Figure 3-12. When aligned, the satellite rotates around the pitch axis and the yaw axis always points to the center of the earth.

The pitch axis (the solar array axis) points in a north-south direction orthogonal to the geostationary orbit plane.

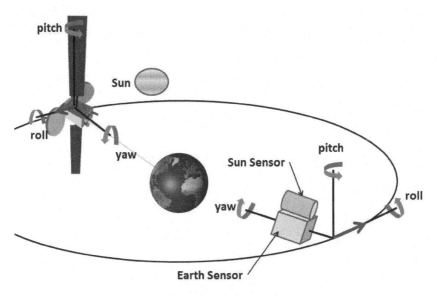

Figure 3-12 Earth and Sun Sensors Mounting

Earth Sensors

Earth sensors used on geostationary orbits measure the difference of radiance of the **earth's limb** (edge of the earth's disc) and space to determine the satellite's pitch and roll attitude. There are different types of earth sensors.

One type that is frequently used on geostationary satellites contains a rotating mirror and fixed mirrors to generate a scanning pattern for an infrared sensor to detect earth to space and space to earth transitions.[10] This type of earth sensor is the size of a shoe box and weighs about 1.5 pounds.

Figure 3-13 shows a scanning pattern with a roll error for this type of earth sensor. The grey pattern is the neutral scan pattern without a roll or pitch error.

The black pattern is shifted downward from the neutral grey scanning pattern that happens with a roll error. With the roll error the black Trace 1 time is shorter than the grey Trace 1 time and the black Trace 2 time is longer than the grey Trace 2 time.

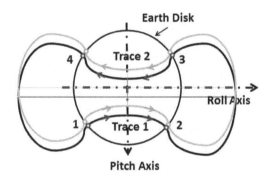

Figure 3-13 Roll Error

Sensor electronics process these measurements to determine the amount of roll error. The roll error information goes to the attitude control electronics which issues commands to change the speed of the appropriate momentum wheel(s). This process continues in a closed loop until the roll error is corrected.

The scanning pattern in Figure 3-14 shows a combined roll and pitch error. Again, the grey pattern is the neutral scan pattern without roll or pitch errors.

The black pattern is shifted down and to the right from the neutral grey scanning pattern. The difference between the grey and black Trace 1 and 2 times measure the roll and pitch errors.

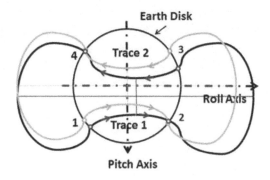

Figure 3-14 Combined Roll and Pitch Error

Sensor electronics process these measurements to determine the amount of roll and pitch errors. The error information goes to the attitude control electronics which issues commands to change the speed of the appropriate momentum wheel(s). This process continues in a closed loop until the errors are corrected.

Sun sensors

The earth sensor measures roll and pitch errors. The **sun sensor** determines yaw errors. The sun sensor is mounted along the satellite's yaw axis and measures the angle between the pitch axis and the sun.

The difference between the measured and predicted angle values is a pointing error that goes to the attitude control electronics.

There are different types of sun sensors. One type that is frequently used on geostationary satellites is a fine sun sensor. The front surface of the sensor is a mirror with slits cut in the reflective metal. Sunlight passes through these slits and then through an optical filter.

Below this is an array of photo sensors interfaced to a micro-controller. The charges on the photo sensors are read by the microcontroller, which processes the image and computes the sun angle. This type of sun sensor is about the size of half a stick of butter and weighs less than a pound.[11]

We discussed that the plane of the earth's equator does not lie in the plane of the earth's orbit or ecliptic plane. The sun moves from 23° below the equatorial plane to 23° above the equatorial plane and back again over the course of a year as shown in Figure 3-15.

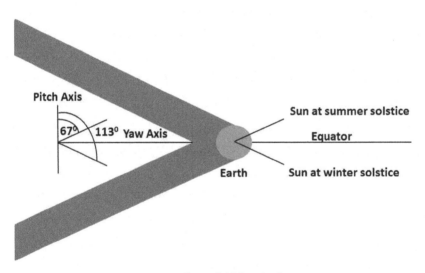

Figure 3-15 Sun Angle

When aligned, the satellite pitch axis is normal (or at a right angle) to the equatorial orbit plane, which makes it in alignment with the earth's axis of rotation. During a year the angle from the pitch axis to the sun varies from 67^0 to 113^0. This angle is 90^0 two times a year at the equinoxes.

We can predict the sun's angles for a perfectly aligned satellite as a function of time. The difference between the measured sun angles and the predicted sun angles determines yaw error.

As discussed before, the satellite is in the earth's shadow around the time of the equinoxes. Also, the sun sensor has its back to the sun for more than half of the orbit because it is mounted along the yaw axis.

Momentum Wheels

Momentum wheels increase the pointing precision and reliability of a satellite and also reduce the amount of thruster propellant required to be carried by the satellite if thrusters alone were used for attitude control. A momentum wheel shown in Figure 3-16 consists of an electric motor attached to a flywheel which when rotated increasingly rapidly causes the satellite to spin the other way in a proportional amount by conservation of angular momentum. The flywheel design has most of its mass in the rim to give maximum momentum exchange capability. The encased wheel assembly mounts inside the satellite.

Figure 3-16 Momentum Wheel (Wikimedia Public Domain)

Satellites use several different momentum wheel configurations. One configuration uses two momentum wheels that have their spin axes oriented in the satellite to provide pointing torque and attitude correction torques around all three satellite axes. Another configuration uses gimbaled (two pivoted rings, capable of swinging freely while mounted to the satellite) momentum wheels.

In this configuration, the gimbals orient the momentum wheel's spin axis in the direction that provides total pointing and correction torques. A highly redundant configuration has momentum wheels oriented along three orthogonal, or right angled, axes as shown in Figure 3-17. The fourth momentum wheel's spin axis is at equal angles to the three orthogonal reference axes.

This fourth momentum wheel provides redundancy in the event that one or more of the primary wheels should fail.

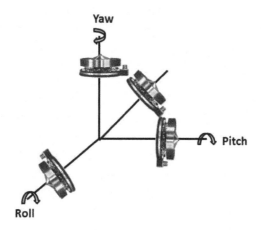

Figure 3-17 Redundant Momentum Wheels

The momentum wheels are as close to the center of mass of the satellite as possible. The attitude control electronics compensates for any offset from the center of mass.

Applying torque around the satellite pitch axis keeps the yaw axis pointing at the center of the earth. In the geostationary orbit, the proper torque results in a uniform manner over 360^0 during the 24 hour orbital period, which corresponds to about a 0.25^0 per minute torque rate.

Attitude sensors report deviations in attitude, caused by external perturbation torques or rotations, to the attitude control electronics which calculates the component torque corrections to make around the reference axes.

To make these corrections, the momentum wheels' speeds are either increased or decreased in the opposite direction to the perturbation torques. When the satellite achieves its desired orientation, it can then halt its rotation by **braking** the momentum wheels by the same amount. Over time momentum wheels build up stored momentum that needs to be removed because the strength of the materials of a momentum wheel establishes a speed at which the wheel could come apart and therefore how much angular momentum it can store. Attitude thruster firings unload the momentum wheels.

Propulsion

How do you make orbit and attitude changes? The satellite propulsion system consists of thrusters and propellant to provide thrust for making velocity change maneuvers for injection into the geostationary orbit and station-keeping to keep the satellite over the equator at a specified longitude. Other propulsion operations include: sun acquisition, earth acquisition, attitude control, desaturation of the momentum wheels, station change (if required), and boost from geostationary orbit at end of life.

A **liquid bipropellant** (fuel and oxidizer) system can satisfy the propulsion requirements with the appropriate selection of thrusters. A typical configuration consists of one high thrust orbit maneuver thruster and twelve low thrust Attitude and Orbit Control (AOC) thrusters shown in Figure 3-18, fuel, oxidizer, and propellant tanks. Most of the propulsion subsystem's components mount on the propulsion module, which is built around the central structural thrust tube. The orbit maneuver thruster mounts on one end of the structure and the twelve AOC thrusters hard-mount to the central structure by brackets and panels for alignment stability. Four pitch and yaw thrusters mount on one end and four others on the other end. Four roll thrusters mount on a side panel. Six of the twelve AOC thrusters are primary with the remaining six acting as backup for complete thruster redundancy.[12]

**Orbit Maneuver
Thruster**

AOC Thruster

Figure 3-18 Thrusters (NASA-Goddard Space Flight Center)

Figure 3-19 is a schematic representation of how the orbit maneuver thruster and AOC thrusters mount on the satellite to perform major orbit maneuvers and attitude and orbit control of the satellite. The AOC thrusters are in opposing, redundant pairs in all three axes.

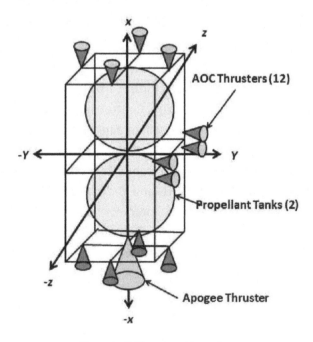

Figure 3-19 Thrusters Mountings

Sample Communications Satellites

Appendix E presents sample communications satellites that are stationed in the geostationary orbit. Table 3-4 shows the mass, power, and number of transponders for these sample communications satellites.

Table 3-4 Sample Satellite Capabilities

Satellite	Launch Mass	Dry Mass	Power	Number of Transponders	
	(pounds)	(pounds)	(watts)	Ku	C
AMC-9	10,985	5,358	10,000	24	
Astra 2C	9,760	5,358	7,000	32	
DirecTV-9S	14,736	6,334	13,900	52	
Echostar 7	10,789	5,894	13,000	32	
Eutelsat W-5	8,493	5,090	5,900	24	
Intelsat 905	12,654	5,316	10,000	22	76
NSS-7	12,056	6,698	3,900	36	36

From Table 3-4, for the first five Ku-band satellites we observe:

- The average launch mass is 10,953 pounds.
- The average dry mass is 5,607 pounds.
- The ratio of launch divided by dry mass is about 2.
- The average dry mass per transponder is 130 pounds.
- The average total power is 9,100 watts.
- The average power per transponder is 310 watts.

Also from Table 3-4, for the last two duel C/Ku-bands satellites we observe:

- The average launch mass is 12,355 pounds.

- The average dry mass is 6,007 pounds.

- The ratio of launch divided by dry mass is about 2.

- The average dry mass per transponder is 74 pounds.

- The average total power is 6,950 watts.

- The average power per transponder is 78 watts

4

Satellite Motion

Satellite motion can be explained by considering the famous 18th Century scientist, Sir Issac Newton's Cannonball experiment shown in Figure 4-1. Imagine there is a cannon on top of a mountain with its barrel aimed horizontal to the earth's surface. Ignore the atmosphere and fire a cannonball. It will travel a certain distance before it falls to earth due to gravity. Fire another cannonball with greater speed than the first and it will go a greater distance than the first before it falls to earth due to gravity.

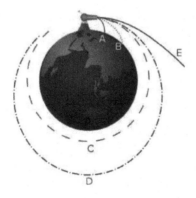

Figure 4-1 Newton's Cannonball (Wikimedia Public Domain)

Add enough speed to the cannonball and it will continue to fall to the earth due to gravity, but it will not hit the earth because it is in orbit around the earth. Continuing to add speed to the cannonball will result in larger oblong or elliptical orbits. If enough speed is added to the cannonball, it will escape earth's gravity and travel through space.

Kepler's Laws

Johannes Kepler, a 17th century German mathematician, astronomer, and astrologer, formulated three laws of planetary motion. They are as follows:

- The path of the planets about the sun is elliptical in shape, with the center of the sun being located at one focus.
- An imaginary line drawn from the center of the sun to the center of the planet will sweep out equal areas in equal intervals of time.
- The ratio of the squares of the periods of any two planets is equal to the ratio of the cubes of their average distances from the sun.

Kepler's first law (referred to as the law of ellipses) states that all planets orbit the sun in a path that resembles an ellipse, with the sun being located at one of the foci of that ellipse.

Kepler's second law (referred to as the law of equal areas) describes the speed at which any given planet will move while orbiting the sun. The speed at which any planet moves through space is constantly changing. A planet moves fastest when it is closest to the sun and slowest when it is furthest from the sun. Yet, if an imaginary line were drawn from the center of the planet to the center of the sun, that line would sweep out the same area in equal periods of time.

Kepler's third law (referred to as the law of harmonies) compares the orbital period and radius of the orbit of a planet to those of other planets. Unlike Kepler's first and second laws that describe the motion characteristics of a single planet, the third law makes a comparison between the motion characteristics of different planets. The comparison being made is that the ratio of the squares of the periods to the cubes of their average distances from the sun is the same for every one of the planets.

Kepler's laws are strictly only valid for a lone zero-mass object orbiting the sun. Nevertheless, Kepler's laws form a useful starting point for calculating the orbits of planets that do not deviate too much from these restrictions.

Elliptical Orbits

Although Kepler's Laws were developed for the solar system planets, they are also valid for satellites.

From Kepler's first law: satellites rotate around the earth in **elliptical orbits** as shown in Figure 4-2, where:

r =   the distance from the center of the earth to the satellite

v =   the speed of the satellite along the orbit path

a =   **the semi-major axis**, one half of the length of the major axis of the ellipse

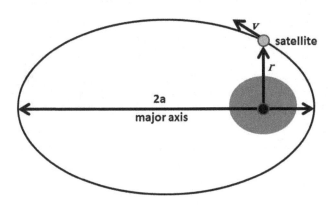

Figure 4-2 Elliptical Orbit

Orbit Equations

Figure 4-3 shows:

At perigee **(P)** the satellite is **closest** to the center of the earth and has **maximum speed**.

At apogee **(A)** the satellite is **farthest** from the center of the earth and has **minimum speed**. Also,

$r_{P=}$ the distance from the center of the earth to perigee

$r_A=$ the distance from the center of the earth to apogee

$v_P=$ the speed of the satellite at perigee

$v_A=$ the speed of the satellite at apogee

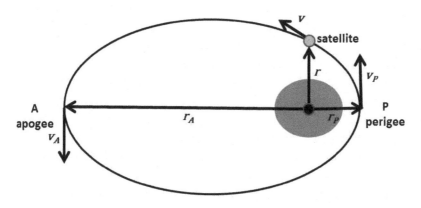

Figure 4-3 Orbit Parameters

The *semi-major axis* is calculated as follows:

$$a = \frac{r_P + r_A}{2}$$

From Kepler's third law, the **period of the orbit** (T), the time it takes for the satellite to make one orbit revolution, is calculated by the equation:

$$T = 2\pi \sqrt{\frac{a^3}{\mu}}$$

Where, $\quad \mu = the\ standard\ universal\ constant$

The speed along the orbit path (v) is calculated by the equation:

$$v = \sqrt{\mu \left[\frac{2}{r} - \frac{1}{a}\right]}$$

For circular orbits: $\qquad\qquad\qquad r = a$

Circular orbit speed is: $\qquad\qquad v = \sqrt{\frac{\mu}{r}}$

The above equations are sufficient to determine the speed of satellites at different altitudes above the earth in circular orbits and the speeds of satellites in elliptical orbits at perigee and apogee.

Determining the distance and speed as a function of time along an elliptical orbital path requires more advanced equations as given in the References[1].

For now we are only considering orbits in the plane of the equator and not orbits that are inclined to the equatorial plane. Later we will consider inclined orbits.

Eccentricity

Figure 4-4 shows orbits with different eccentricities. The more the orbit is stretched out or the larger r_A is to r_P, the greater the eccentricity.

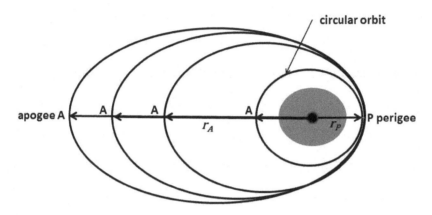

Figure 4-4 Eccentricity

Eccentricity:
$$\epsilon = \frac{r_A}{a} - 1$$
$$\epsilon < 1$$

For circular orbits:
$$r_P = r_A = a$$
$$\epsilon = 0$$

Pure circular satellite orbits are not physically realizable due to space environmental perturbations and earth's anomalies.

Simplifying Calculations

The orbital period and speed equations only require basic mathematics of raising the semi-major axis to the third power and taking square-roots. The thing that makes it computationally complicated is the value of the **Standard Universal Constant**, which in scientific notation is given as:

$$\mu = 1.23936 \: X \: 10^{12} \: mi^3/hr^2$$

This constant is usually measured in metric units, but I've shown it here in miles and hours, because these units are more familiar to most Americans. The 10^{12} means that the decimal point of the number has been moved 12 places to the left meaning that the size of the number is 1.239+ trillion. This size number can easily be handled in a computer, but we're going for simplification.

With the method to follow, you don't even have to remember the value of this constant. If you normalize distances and speeds with respect to the earth's radius and the speed of a *Surface Circular Satellite*, calculations will be relatively easy. Appendix E presents the earth's parameters.

Consider an imaginary satellite that is in a circular orbit at the earth's surface as represented in Figure 4-5. This *Surface Circular Satellite* has a semi-major axis equal to the radius of the earth.

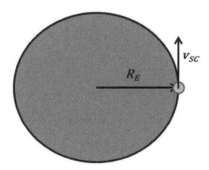

Figure 4-5 Surface Circular Satellite

Orbit parameters of the *Surface Circular Satellite* are as follows:

Earth's Radius: R_E = 3,963 miles

Surface Circular
Satellite Speed: $v_{SC} = \sqrt{\dfrac{\mu}{R_E}}$ = 17,684 mph

Surface Circular
Satellite Period: $T_{SC} = 2\pi \sqrt{\dfrac{R_E{}^3}{\mu}}$ = 1.41 hours

Normalizing orbit parameters by dividing by the *Surface Circular Satellite* parameters are as follows:

Normalized Distance: $r' = \dfrac{r}{R_E} = \dfrac{r}{3,963}$

Normalized
Semi-Major Axis: $a' = \dfrac{a}{R_E} = \dfrac{a}{3,963}$

Normalized Speed: $v' = \dfrac{v}{v_{SC}} = \dfrac{v}{17,684}$

Normalized
Orbital Period (hours): $T' = \dfrac{T}{T_{SC}} = \dfrac{T}{1.41}$

Distance (miles): $r = R_E r'$

Speed (mph): $v = v_{SC} \sqrt{\left[\dfrac{2}{r'} - \dfrac{1}{a'}\right]}$

Orbital Period (hours): $T = T_{SC} \sqrt{a'^3}$

5

Mission Planning Example

The Mission

The mission is to place a hypothetical communications satellite into a geostationary orbit at a 90^0 West longitude station above the earth's equator and keep it on station.

Mission Planning involves many factors. For our example, we shall limit these factors to determining: transfer orbit parameters, required speed corrections, transfer times, and propellant mass requirements. We shall determine mission planning parameters for transferring a stowed communications satellite from a 200 miles altitude circular equatorial parking orbit into a circular geostationary orbit and station it at a 90^0 West longitude above the equator as shown in Figure 5-1. The primary outputs of this step are the speed correction and time to inject into the transfer orbit, and the speed correction to inject into geostationary orbit.

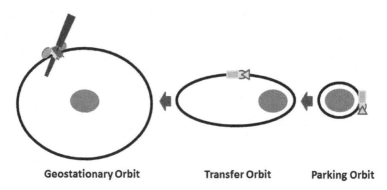

Geostationary Orbit   Transfer Orbit   Parking Orbit

Figure 5-1 Mission

In the next chapter, we refine mission planning with launch orbit inclination and correction, multiple transfer orbits, and inclination and longitude drift.

Parking Orbit

Looking down from space at the North Pole, Figure 5-2 shows the circular parking orbit.

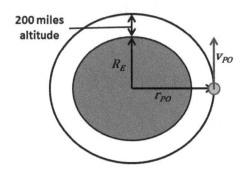

Figure 5-2 Parking Orbit

Table 5-1 lists the parameters for this orbit.

Table 5-1 Circular Parking Orbit Parameters

Parameter	Equation	Value
Earth's Radius:	R_E	3,963 miles
Parking Orbit Radius:	$r_{PO} = R_E + 200$ miles	4,163 miles
Normalized Radius:	r'_{PO}	1.05
Speed:	$v = v_{SC}\sqrt{\dfrac{1}{r'_{PO}}}$	17,253 mph

Geostationary Orbit

Looking down from space at the North Pole, Figure 5-3 shows the circular geostationary orbit.

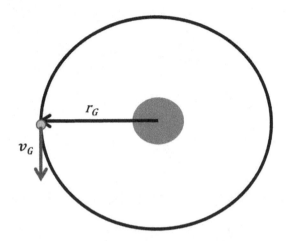

Figure 5-3 Geostationary Orbit

Table 5-2 lists the orbit parameters for the geostationary orbit.

Table 5-2 Geostationary Orbit Parameters

Parameter	Equation	Value
Earth's Radius:	R_E	3,963 miles
Geostationary Radius:	r_G	26,200 miles
Normalized GEO Radius:	$r'_G = r_G/R_E$	6.61
Geostationary Speed:	$v_G = v_{SC}\sqrt{\dfrac{1}{r'_G}}$	6,878 mph
Semi-Major Axis:	$a' = r'_G$	6.61
Orbital Period:	$T = T_{SC}\sqrt{a'^3}$	23.93 hours

The period of the geostationary orbit is slightly less than 24 hours to account for earth's rotation around the sun.

Transfer Orbit

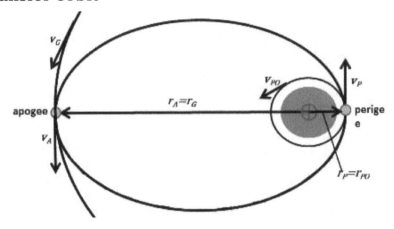

Figure 5-4 Transfer Orbit

Table 5-3 lists the transfer orbit parameters shown in Figure 5-4.

Table 5-3 Transfer Orbit Parameters

Parameter	Equation	Value
Normalized Perigee:	$r'_P = r'_{PO}$	1.05
Normalized Apogee:	$r'_A = r'_G$	6.61
Normalized Semi-Major Axis:	$a' = \frac{1}{2}[r'_P + r'_A]$	3.83
Eccentricity	$\epsilon = r'_A/a' - 1$	0.73
Perigee Speed:	$v_P = v_{SC}\sqrt{\left[\frac{2}{r'_P} - \frac{1}{a'}\right]}$	22,675 mph
Apogee Speed:	$v_A = v_{SC}\sqrt{\left[\frac{2}{r'_A} - \frac{1}{a'}\right]}$	3,658 mph
Orbital Period:	$T = T_{SC}\sqrt{a'^3}$	10.64 hours
Transfer Time:	$TT = \frac{1}{2}T$	5.32 hours

Transfer Orbit Injection

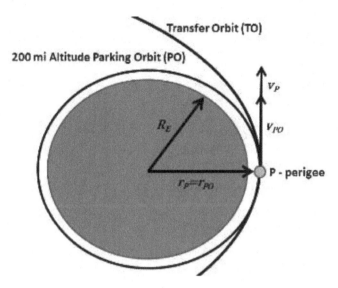

Figure 5-5 Transfer Orbit Injection

Looking down from space at the North Pole, Figure 5-5 shows that the satellite injects into the transfer orbit at perigee by increasing the speed along the orbit path as follows:

$$v_p = 22{,}675 \ mph$$

$$v_{PO} = 17{,}253 \ mph$$

The change in velocity or speed is defined as **Delta-V, ΔV**, therefore:

$$\Delta V_P = v_P - v_{PO} = 5{,}422 \ mph$$

Geostationary Orbit Injection

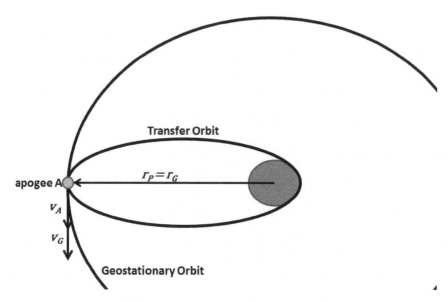

Figure 5-6 Geostationary Orbit Injection

After 5.32 hours on the transfer orbit, the satellite arrives at the geostationary orbit altitude at apogee. Looking down from space at the North Pole, Figure 5-6 shows the satellite is injected into the geostationary orbit by increasing the speed along the orbital path as follows:

$$v_G = 6,878\ mph$$

$$v_A = 3,658\ mph$$

$$\Delta V_A = v_G - v_A = 3,219\ mph$$

Achieving Longitude Station

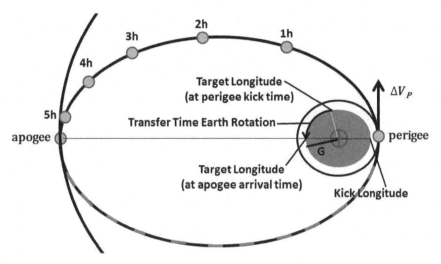

Figure 5-7 Achieving Longitude Station

We want the satellite to arrive at the geostationary orbit altitude when it is over the 90^0 West target longitude as shown in Figure 5-7. The earth rotates at 15^0 per hour while the satellite is on the Transfer Orbit and the Transfer Time is 5.32 hours, therefore, the earth rotates about 80^0 during this time.

The longitude where we want to make the perigee kick maneuver is 100^0 west of our Target Longitude, which corresponds to 170^0 East Longitude.

From Kepler's second law, the satellite reaches maximum speed at perigee and decreases along the transfer orbit because the satellite sweeps out an equal area of the ellipse in an equal time. Figure 5-7 shows the approximate position of the satellite on the transfer orbit each hour after the kick maneuver at perigee.

Orbits Summary

Table 5-4 summarizes our three orbits. We now have a sense for large distances and high speeds that satellites need to stay in orbit. Compared to 65 mph freeway speeds, the slowest speed we calculated is 56 times faster, while the highest speed is 349 times faster. A commercial airliner's speed is about 550 mph which gives a 6.7 to 41 times range. Compared to the speed of sound (768 mph), the range is between Mach 4.8 and Mach 29.5.

Table 5-4 Orbits Summary

Orbit	a	T	r	Altitude	v	ΔV
	miles	hours	miles	miles	mph	mph
Parking (200 mi)	4,163	1.52	4,163	200	17,253	
Transfer (perigee)	15,182	10.64	4,163	200	22,675	5,422
Transfer (apogee)	15,182	10.64	26,200	22,236	3,658	3,220
GEO	26,200	23.93	26,200	22,236	6,878	
Transfer Time = 5.32 hours						

The distance to geostationary orbit is almost the same as the distance around the earth's circumference. It also is equivalent to 10 trips from coast to coast. The time it takes to get to geostationary orbit is almost the same time it takes to fly coast to coast.

The Surface Circular Satellite period is 84 minutes. The 200 mile altitude circular orbit's period is about 91 minutes, which is about the same duration of a feature film. In our Transfer Orbit the perigee Delta-V is more than 1.5 times the apogee Delta-V. The transfer orbit has an eccentricity of 0.73.

MISSION PLANNING EXAMPLE

6

Mission Planning Refinement

In the previous chapter, we determined that a 3,220 mph Delta-V is needed to inject into the geostationary orbit from our transfer orbit. We also determined that we needed to inject into the transfer orbit at 170^0 East longitude and 200 miles altitude above the earth in order to arrive at the 90^0 West longitude station in the geostationary orbit. In this chapter, we will refine mission planning by considering orbit inclination, multiple transfer orbits, and station-keeping Delta-V requirements.

Currently, communications satellites directly launch into geostationary orbit from the ground and not via a parking orbit like some were before the Space Shuttle was discontinued. For our hypothetical mission planning example, we will assume that the launch vehicle injected our satellite into a transfer orbit that resulted in the same 3,220 mph Delta-V to inject into the geostationary orbit.

Satellites launch due East to take advantage of the earth's rotation. The higher the launch site latitude, the greater the orbit inclination angle. Removing the inclination angle to achieve an equatorial orbit requires propellant. The most efficient time to make the inclination corrections is at apogee in combination with the geostationary orbit injection maneuver.

After the multiple transfer orbits, we deploy the antennas and solar arrays and align and position the satellite on station.

Orbital perturbations caused by gravitational attractions of the sun, moon and earth anomalies cause drifts in orbit inclination and longitude. The satellite stays on station by station-keeping thruster maneuvers.

Thrusters and propellant provide thrust for apogee maneuvers, station-keeping, sun acquisition, earth acquisition, attitude control, momentum wheels desaturation, and boost from geostationary orbit at end of life. We derive the maneuver propellant mass required in the next chapter.

Orbit Inclination

Launch Site Latitude

Except for China, most geostationary satellites launch from Cape Canaveral, Florida; Kourou, French Guiana; or Baikonur, Kazakhstan. They arrive at the geostationary orbit altitude at apogee of an inclined transfer orbit. The satellite's propulsion system provides Delta-V maneuvers for inclination removal and increases the satellite's speed by the amount necessary to get in the geostationary orbit.

Figure 6-1 shows that launching a satellite from Cape Canaveral (28.5^0 latitude) results in an inclined orbit to the equatorial orbit plane of 28.5^0. Likewise, launching a satellite from Kourou, French Guiana (5^0 latitude) or Baikonur, Russia (45.6^0 latitude) results in an inclined orbit to the equatorial orbit plane of 5^0 or 45.6^0, respectively. The orbit inclination is removed at either the ascending or descending nodes.

The amount of Delta-V to remove the inclination (i) but keep the same speed (v) before and after the maneuver is determined by the following equation:

$$\Delta v = 2v \sin \frac{i}{2}$$

We need to remove any inclination angle due to launch site latitude between our parking orbit and the equator. We could make an inclination maneuver before we made our transfer orbit perigee maneuver or we could remove the inclination when we reached apogee.

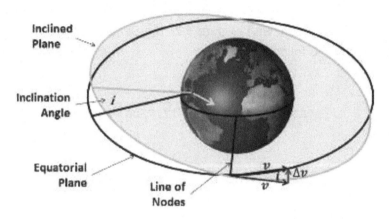

Figure 6-1 Launch Orbit Inclination

Because the transfer orbit speed at perigee is much greater than at apogee, it is more efficient to make the inclination removal maneuver at apogee. The Delta-V for inclination corrections at perigee and apogee are shown in Table 6-1. The Delta-V required at perigee to remove 5^0 is 4.72 times what it is at apogee.

Table 6-1 Inclination Delta-V Corrections Comparison

Launch Site	Inclination (degrees)	Perigee (mph)	Apogee (mph)
Kourou, French Guiana	5	1,505	319
Cape Canaveral, Florida	28.5	8,494	1,801
Baikonur, Kazakhstan	45.6	13,372	2,836

Apogee Inclination Angle Correction

For the case of transfer to a geostationary orbit from an inclined orbit, it is much more efficient to combine the inclination correction with the apogee injection maneuver as shown in Figure 6-2. The Delta-V for this combined maneuver is given as follows:

$$\Delta v = \sqrt{v_A{}^2 + v_G{}^2 - 2v_A v_G \cos i}$$

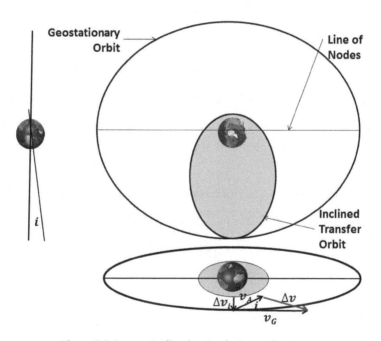

Figure 6-2 Apogee Inclination Angle Correction

Table 6-2 shows the Delta-V required for making inclination change maneuvers at apogee and in combination with the apogee injection maneuver. There is a clear advantage to making a combined apogee maneuver and the advantage increases with increasing inclination angle.

Table 6-2 Combined Inclination Delta-V Corrections

Launch Site	Inclination (degrees)	Apogee (mph)	Combined* (mph)
Kourou, French Guiana	5	319	29
Cape Canaveral, Florida	28.5	1,801	837
Baikonur, Kazakhstan	45.6	2,836	1,828
*Combined Apogee Correction – Zero Inclination Correction (3,220 mph)			

Multiple Transfer Orbits

Figure 6-3 shows a direct launch to geostationary orbit from Kourou, French Guiana using multiple transfer orbits. The necessary steps to ready the satellite for its communication mission include:

1. Launch and Ascent
2. Reorientation / Apogee Maneuvers / Remove Inclination
3. Reflector Antennas Deployment
4. Despin Maneuver / Start Solar Panel Deployment
5. Solar Panel Deployment Complete
6. Sun Acquisition / OMNI Repositioning
7. Earth Acquisition

We use multiple transfer orbits to nudge the satellite into position and carefully prepare the satellite for its communications mission.

Figure 6-3 shows four orbit corrections at apogee that sequentially raise the perigee to the geostationary orbit altitude.

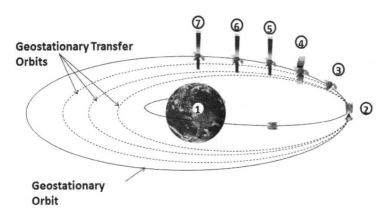

Figure 6-3 Multiple Transfer Orbits

We know from our previous calculations summarized in Table 5-4 Orbits Summary: the transfer orbit parameters of apogee speed, apogee distance (equal to geostationary orbit radius), and total Delta-V to inject into geostationary orbit. We will use 25% of the total Delta-V to generate four equal speed corrections resulting in intermediate transfer orbits.

After adding the correction to the current apogee speed, we calculate the semi-major axis from the speed equation. The new perigee distance is twice the semi-major axis minus the geostationary orbit radius. Continuing this process three more times results in establishing the desired geostationary orbit. The amount of time spent in each of these orbits varies.

Derivation of the normalized semi-major axis from the speed equation is as follows:

The Speed Equation:
$$v = \sqrt{\mu \left[\frac{2}{r} - \frac{1}{a} \right]}$$

Normalized Speed:
$$v' = \sqrt{\left[\frac{2}{r'} - \frac{1}{a'} \right]}$$

Squaring:
$$v'^2 = \frac{2}{r'} - \frac{1}{a'}$$

Normalized Semi-Major Axis:
$$a' = \frac{1}{\left[\frac{2}{r'} - v'^2 \right]}$$

The values we need to calculate the parameters of the multiple transfer orbits are presented in Table 6-3 on the next page.

To each of the multiple transfer orbits, we add one quarter of the total Delta-V of 3,220 mph to get the satellite in a geostationary orbit at apogee. The four speed corrections will each be 805 mph. From our previous calculations, the initial apogee speed is 3,658 mph. Adding 805 mph to each of the four speed maneuvers results in the apogee speeds listed in Table 6-4 on the next page.

Table 6-3 Multiple Transfer Orbits Parameters

Parameter	Symbol	Value	Symbol	Normalized Value
Surface Circular Orbit				
Earth's Radius:	R_E	3,963 miles	Base	1.0000
Speed:	v_{SC}	17,684 mph	Base	1.0000
Transfer Orbit				
Apogee Distance:	r_A	26,200 miles	r'_A	6.6100
Apogee Speed:	v_A	3,658 mph	v'_A	0.2069
Delta Speed to GEO:	ΔV_A	3,220 mph	$\Delta V'_A$	0.1821
Geostationary Orbit				
Radius:	r_G	26,200 miles	r'_G	6.6100
Speed:	v_G	6,878 mph	v'_G	0.3890

Table 6-4 Multiple Delta-V Maneuvers

Speed Maneuver	ΔV	v_A	v'_A
	mph	mph	normalized
Initial		3,658	0.2069
1	805	4,464	0.2524
2	805	5,269	0.2980
3	805	6,074	0.3435
4	805	6,878	0.3890

Next, we calculate the semi-major axis, perigee distance and period for the four intermediate transfer orbits. The apogee for each of the orbits is the same, 26,200 miles.

Semi-Major Axis:
$$a' = \frac{1}{\left[\frac{2}{r'} - v'^2\right]} \qquad a = R_E a'$$

The Perigee Distance:
$$r'_P = 2a' - r'_G \qquad r_P = R_E r'_P$$

The Orbital Period:
$$T' = \sqrt{a'^3} \qquad T = T_{SC} T'$$

The results of our calculations are shown in Table 6-5

Table 6-5 Multiple Orbits Parameters

Orbit	a	v_A	r_P	T
	miles	mph	miles	hours
initial	15,182	3,659	4,163	10.64
1	16,592	4,464	6,984	12.08
2	18,536	5,269	10,872	14.26
3	21,468	6,074	16,736	17.78
4	26,200	6,878	26,200	23.93

We have determined the orbit parameters and speed corrections for the injection maneuvers. Next we will determine the longitude to arrive at geostationary orbit and the longitudes to perform apogee maneuvers for each of the multiple transfer orbits.

We use orbital period information for determining what longitude the satellite should arrive at the geostationary orbit to achieve the specified longitude station (90°W for our example). Also, because the apogee maneuvers are controlled from ground stations, it is important to know at what longitudes the apogee maneuvers need to be performed.

We work backwards from the 4th orbit (geostationary orbit) where we want the satellite to be stationed at 90°W to determine **Delta⁰**, the number of degrees the earth rotates (at a $15°$ per hour rotation rate) during each transfer orbits' period, or:

$$Delta^0 = 15^0(T_{GEO} - T_i)$$

Where,

T_i = Transfer orbit period of the i-th transfer orbit

T_{GEO} = Geostationary orbit period

Table 6-6 shows that the satellite needs to initially arrive at the geostationary orbit over a longitude of 74°W to achieve a final station over a longitude of 90°W.

Table 6-6 Multiple Transfer Orbits Apogee Maneuver Longitudes

Orbit	T	T_{GEO} - T	Delta⁰	Apogee Longitude	
	hours	hours	degrees	degrees	
initial	10.64	13	199	74	W
1	12.08	12	178	125	E
2	14.26	10	145	57	W
3	17.78	6	92	88	E
4	23.93			90	W

MISSION PLANNING REFINEMENT

Figure 6-4 shows the multiple transfer orbits for our mission planning example looking down at the North Pole.

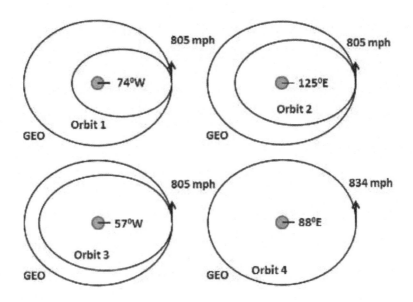

Figure 6-4 Multiple Transfer Orbits Top View

In summary, we divided the required Delta-V into four equal increments of 805 mph, determined the multiple transfer orbits' periods, and determined the starting apogee longitude to achieve a satellite station of 90^0W longitude.

We removed the orbit inclination during the final transfer orbit to minimize interference and collision with other satellites.

The longitudes of other satellites need to be examined to determine if there is a risk of collisions. Also, the transfer orbits' design should include an analysis of debris to avoid collisions.

The injection apogee maneuvers occur at widely separated longitudes where they are commanded by ground stations. If ground stations are not available from these longitudes, the transfer orbits need to be redesigned.

SATELLITE BASICS FOR EVERYONE   89

Station-Keeping

How do we keep the satellite on station? Once on station, the satellite must frequently perform a variety of station-keeping maneuvers over its mission life to compensate for orbital perturbations. These perturbations cause the geostationary orbit to drift in inclination resulting in a North-South shift of the communication earth coverage patterns. Also, they cause an east-west (longitude) drift in the satellite's position above the earth.

Inclination Drift

The principal perturbation is the combined gravitational attractions of the sun and moon, which causes the orbital inclination, shown in Figure 6-5 to increase by nearly one degree per year.

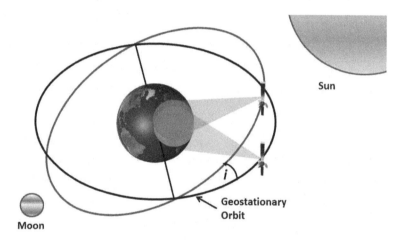

Figure 6-5 Inclination Drift

This perturbation is compensated for by a north-south station-keeping maneuver approximately once every two weeks to keep the satellite within 0.05^0 of the equatorial plane. The average annual velocity increment is about 164 feet/second.

Longitude Drift

The earth anomalies at the equator cause the geostationary satellite to drift in a longitudinal direction. Figure 6-6 shows there are two stable orbit points and two unstable points.

Geostationary satellites will drift toward a stable point and away from an unstable point. Some of the World's major cities are shown in relation to these stability points.

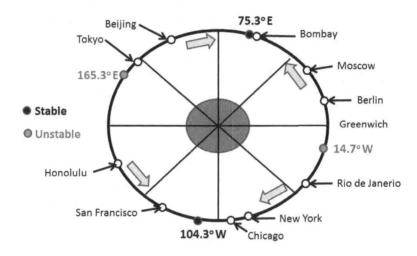

Figure 6-6 Longitude Drift

The longitudinal drift is compensated by east-west station-keeping maneuvers about once a week, with an annual velocity increment of less than 7 feet/second, to keep the satellite within 0.05^0 of its assigned longitude.

It appears that Bombay, India is very close to a stable point and Tokyo, Japan is very close to an unstable point. Chicago, New York, and San Francisco are relatively close to a stable point.

Other perturbations of solar radiation pressure and infrared radiation from the sun in the form of electromagnetic waves flatten (increase an orbit's eccentricity) the orbit and disturbs the orientation of the satellite. A control maneuver that is combined with east-west station-keeping compensates for orbit eccentricity.

We determined the orbit parameters and speed corrections for the injection maneuvers, station-keeping, and attitude control. In the next chapter, we will estimate the required propellant.

7

Propellant Requirements

The satellite propulsion system consists of thrusters, tanks, fuel lines, and propellant to provide thrust for making velocity change (apogee) maneuvers for injection into the geostationary orbit and station-keeping to keep the satellite over the equator at a specified longitude. Other propulsion operations include: sun acquisition, earth acquisition, attitude control during apogee thruster firing, desaturation of the momentum wheels, station change, and boost from geostationary orbit at end of life.

Rocket Equation

Figure 7-1 shows a schematic of a rocket thruster.

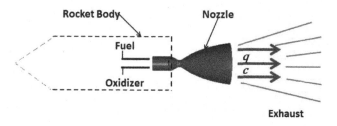

Figure 7-1 Thruster Schematic

The rocket equation in a space vacuum is:

$$F = qc$$

Where,
$$F = \text{thrust}$$
$$q = \text{mass flow}$$
$$c = \text{effective exhaust velocity}$$

For a constant thrust, the **specific impulse** is:

$$I_{sp} = \frac{c}{g}$$

Where, g = acceleration of gravity at the earth's surface

Given the specific impulse, c, is:

$$c = gI_{sp}$$

We can use a simplified approach to determine the amount of propellant necessary to perform the propulsion operations. The change in velocity, called Delta-V, for an operation is given by the equation:

$$\Delta V = c\, ln\frac{m_0}{m_1}$$

Where,

c = effective exhaust velocity

ln = natural log operation

m_0 = initial mass

m_1 = final mass

Solving for m_1,

$$m_1 = m_0 e^{-\left(\frac{\Delta V}{c}\right)}$$

The amount of propellant used for each Delta-V maneuver is:

$$\Delta M = m_0 - m_1$$

The time it takes to achieve the Delta-V or **firing time** is:

$$t = \frac{m_0}{q}\left[1 - \frac{1}{e^{(\Delta V/c)}}\right]$$

Orbit Insertion Maneuvers

The Delta-V for inserting the satellite into the geostationary orbit calculated above is repeated in Table 7-1 with an addition in Delta-V to remove launch inclination during the fourth apogee maneuver.

Table 7-1 Apogee Delta-V Maneuvers

Apogee Maneuver	Delta-V (mph)	Delta-V (feet/second)
1	805	1,181
2	805	1,181
3	805	1,181
4	834*	1,224
*A value of 29 mph is combined with this maneuver to remove the 5^0 orbit inclination that resulted from a Kourou, French Guiana launch.		

Calculations are made using the International System of Units (SI) and then transformed to the U.S. English units.

Force or Thrust, is measured in the International System of Units (SI) as the Newton (symbol: N), and represents the amount needed to accelerate 1 kilogram of mass at the rate of 1 meter per second per second. The standard of mass is a 1-kilogram platinum-iridium cylinder kept in a vault in France.

Force or Thrust is measured in "pounds of force" lbf in U.S. English units. A pound of force is how much force it would take to keep a one-pound object unmoving against the force of gravity on earth.

Please see Appendix A – Unit Conversions for abbreviations and conversions of units.

Our hypothetical example propulsion system consists of a 110-pound apogee thruster and an array of twelve 5-pound thrusters for station-keeping, momentum wheel desaturation, attitude control, and corrections of perturbations. The propellant has a specific Impulse of 305 seconds. The effective exhaust velocity is calculated as follows.

Specific Impulse:	I_{sp}	305 s	305 s
Acceleration of gravity:	g	9.81 m/s²	32.17 ft/s²
Effective Exhaust Velocity:	$c = g\,I_{sp}$	2,991 m/s	9,812 ft/s

Assume satellite launch mass of 3,402 kilograms or 7,500 pounds.

Calculating the propellant mass for apogee maneuver 1:

Initial Mass:	m_0	3,402 kg	7,500 lbs.
Exhaust Velocity:	c	2,991 m/s	9,812 ft/s
Delta-V:	$\Delta V = c\,\ln\dfrac{m_0}{m_1}$	360 m/s	1,181 ft/s
Final Mass:	$m_1 = m_0 e^{-\left(\frac{\Delta V}{c}\right)}$	3016 kg	6,650 lbs.
Delta-Mass	$\Delta M = m_0 - m_1$	385 kg	850 lbs.

Calculating the firing time of apogee maneuver 1:

Thrust:	$F = qc$	489 N	110 lbf
Mass Flow:	$q = \dfrac{F}{c}$	0.16 kg/s	0.36 lbs./s
Firing Time:	$t = \dfrac{m_0}{q}\left[1 - \dfrac{1}{e^{(\Delta V/c)}}\right]$	32 minutes	32 minutes

PROPELLANT REQUIREMENTS

The results of these calculations for all four of the Delta-V maneuvers are shown in the following tables.

Table 7-2 shows the delta-mass for the four apogee maneuvers.

Table 7-2 Apogee Delta-Mass

Apogee Maneuver	Delta-V (feet/second)	Initial Mass (pounds)	Final Mass (pounds)	Delta-Mass (pounds)
1	1,181	7,500	6,650	850
2	1,181	6,650	5,897	793
3	1,181	5,897	5,228	669
4	1,224	5,228	4,616	612
Total	4,767			2,924

Table 7-3 adds the firing times to the final results.

Table 7-3 Apogee Summary

Apogee Maneuver	Delta-V (feet/second)	Delta-M (pounds)	Firing Time (minutes)
1	1,181	850	32
2	1,181	793	29
3	1,181	669	25
4	1,224	612	23
Total	4,767	2,924	109

Station-Keeping

As discussed above, perturbations due to the pull of the moon and sun and irregularities of the earth at the Equator cause the satellite to drift from its assigned station. The perturbations from the moon and sun cause an annual orbit inclination drift of about 0.85 degrees. The inclination drift causes a North-South motion of the communication beam coverage pattern.

A Delta-V of 164 feet/second is required per year to eliminate this drift and keep the satellite on station.

For a 15 year satellite lifetime, the Delta-V is 2,460 feet/second. Using 5 pound thrusters and a specific impulse of 305 seconds, these maneuvers consume about 1,016 pounds of propellant.

The irregularities of the earth at the equator cause a satellite East-West drift that requires about 7 feet/second correction per year. For a 15 year operational lifetime, the requirement is 98 feet/second which corresponds to 46 pounds of propellant with 5 pound thrusters and a specific impulse of 305 seconds.

Attitude Control

Over time momentum wheels build up stored momentum that needs to be removed because the strength of the materials of a momentum wheel establishes a speed at which the wheel could come apart and therefore affect how much angular momentum it can store. Attitude thrusters unload the momentum wheels.

Momentum wheel desaturation, sensors acquisitions and attitude corrections for perturbations from solar radiation pressure, meteorite impacts and other forces require about 300 feet/second over 15 years which corresponds to 138 pounds of propellant with 5 pound thrusters and a specific impulse of 305 seconds.

Propellant Mass Summary

Satellite Propellant

Table 7-4 shows the propellant mass summary. The propellant allocated to injection into the geostationary orbit is about 70% of the total propellant. Of the remaining 30% on-orbit propellant, about 75% is for correcting the north-south latitude inclination drift.

Table 7-4 Propellant Mass Summary

Maneuver	Delta-V (feet/second)	Thruster (pounds)	Propellant (pounds)
Injection	4,767	110	2,924
On-Orbit	2,965	5	1,360
N/S	2,460	5	1,016
E/W	98	5	46
Other	300	5	138
Total	7,732		4,284

Launch Propellant

We assumed an Ariane 5 launch from the Centre Spatial Guyanais at Kourou in French Guiana, where the proximity to the equator gives a significant launch advantage because of the 5^0 latitude. The Ariane 5 launch vehicle has successfully launched multiple satellites.

Ariane 5 can launch 23,143 pounds of satellite to geostationary orbit with about 1.5 million pounds of propellant. Proportioning the total 1.5 million pounds of launch propellant to our 7,500 pounds satellite results in about one half of a million pounds of launch propellant.

Total Mass Summary

Table 7-5 presents a total mass summary. At the start of injection into geostationary orbit, we assumed a total satellite mass of 7,500 pounds. This mass includes the mission payload, bus and propellant weight. We calculated a total propellant mass of 4,284 pounds. Adding a 5% (227 pounds) propellant reserve and 50 pounds of propellant for repositioning the satellite out of geostationary orbit at end-of-life (EOL), our total propellant mass is 4,561 pounds.

Table 7-5 Total Mass Summary

Event	Mass (pounds)	
Start Injection into Geostationary Orbit (GEO)	7,500	
Injection Delta- V Propellant		2,924
Beginning of Orbit Life (BOL)	4,576	
On-Orbit Delta- V Propellant		1,360
End of Orbit Life (EOL)	3,216	
5% Reserve Delta- V Propellant		227
De-Orbit Delta- V Propellant		50
Dry Satellite	2,939	
Total Propellant		4,561

Subtracting the total propellant mass from the total satellite mass equates to a dry satellite mass of 2,939 pounds. The ratio of the total satellite mass to the dry satellite mass is about 2.5. This mass ratio is useful for ballpark estimates to determine the amount of total propellant required for a specific dry satellite mass for missions like our hypothetical example.

Mission Optimization

We now have ballpark values for our mission. This is a starting point for analyses, trade studies, and simulations to design the best possible cost-effective system and mission.

In our mission planning example, we did not take into account the other satellites in the geostationary orbit zone when we were making our orbit maneuvers. A careful analysis would help avoid interference and collisions.

Trade studies would focus on maximizing payload capabilities, performance, and orbital lifetime versus propellant mass.

Detailed orbit analyses and simulations would include perturbations and anomalies and provide improved estimates of orbital parameters.

Error analyses would provide refined estimates of pointing and control parameters.

Launch vehicle, launch site, and logistics trade studies would determine the best satellite orbit delivery system.

Analyses of operations and maintenance plans and procedures would help to decrease operational costs.

Mission payload capability and performance would be maximized and its mass decreased with the latest technology. By injecting new technology while keeping the same or better performance, the satellite bus components' mass and power consumption would be minimized to allow increased mission payload mass or more propellant for longer lifetime.

Reliability and redundancy analyses would help to increase lifetime. Communications service outages can be costly and bad for business. Failure analyses and simulations would provide contingency plans.

In summary, all physically realizable alternatives would be considered with the objectives of highest possible communications capabilities, performance, and lifetime with the lowest possible cost.

Like all businesses, satellite customers and service providers always want "the biggest bang for the buck."

8

Conclusions

Satellites have become very important parts of our lives. We depend on them for communications services. They provide weather, earth resources, and sea state information. GPS provides accurate position information so we can navigate to a destination. We can identify and count objects from satellites. They detect distress signals for emergency rescues. Tracking tagged wildlife provides migration information.

Satellites are critical to our security and law enforcement. Reconnaissance satellites uncover enemy forces, equipment, supplies, bases, and movements. They detect illegal crops and smuggling routes. Communication satellites provide global command and control of our military assets. With satellites, we can remotely fly drones over war zones from bases in the United States.

Geostationary communications satellites serve television, radio, commercial business, Internet, and telephone users over a large coverage area on the land, in the air, and on the sea.

Teleports access satellites to provide customer communication services that are widely distributed to provide global coverage of services. They are responsible for processing and routing the signals they transmit and receive from the satellite and terrestrial communications segments carried on land lines, land cable, and submarine cable.

Geostationary communications satellites consist of a communications payload of antennas and a repeater and a satellite bus, which supplies electrical power, performs orbit maneuvers, keeps the satellite on station, maintains satellite attitude and points the satellite in the right direction.

Station-keeping compensates for orbit inclination drift and east-west longitude drift.

Attitude control keeps the communications coverage area in the same place and compensates for rotational perturbations.

Satellite lifetime mostly depends on the amount of propellant remaining to perform station-keeping and attitude control.

On average it takes about 50 times the satellite weight in propellant to launch a satellite into geostationary orbit. More than half of the satellite's weight is propellant used for injection into geostationary orbit, station-keeping and attitude control.

At end of their lives, we need to reposition geostationary satellites to higher orbits to make way for their replacements in specific longitude slots away from the geostationary orbit zone to minimize the risk of collision with active satellites, with themselves and with space debris. Also, some satellites are only a few miles apart, which create a risk of collisions in the event of a hardware malfunction or a position control error. Any of these collisions would create thousands of more pieces of space debris. The space debris problem will continue to accelerate and some scientists are predicting a possible chain reaction.

Costs of satellite ownership include: the purchase price of the satellite, launch costs, insurance, and operating costs. Insurance is a significant part of the cost, so satellite owners look for reliable satellites.

We've come a long way since Sputnik. Satellite technology has advanced exponentially. The evolution of satellite payloads, including: optical sensors, audio sensors, radars, infrared sensors, ultra violet sensors, telecommunications systems, scientific laboratories, and weapons continue to be remarkable.

To keep pace with the payload advances the satellite bus components continue to improve. Earth, sun and star sensors will continue to improve with solid-state designs. Solar cell technology advances will provide higher efficiency and reliability in power generation.

In the future we will see a steady transition to ion propulsion. The improvements in fuel efficiency permit the savings in mass to be used for increasing the revenue-generating payloads (with attendant increase in solar arrays, batteries and thermal control systems to power them), extending the lifetimes in orbit, or reducing the satellite weight to permit a more economical launch vehicle.

Their operational lifetime would increase significantly if satellites could be repaired and refueled in-orbit. NASA states: "The technology and tools already exist to allow people and robots to repair and refuel satellites in orbit. The fundamental objective is to prove that you don't have to design your satellites to be refueled in-orbit, you can refuel existing fleets." NASA plans to demonstrate in-orbit satellite refueling at the international space station with the help of a robot.[13]

Someday we will use vehicles with plasma rockets to perform space maintenance—fixing, repositioning, refueling, and clearing out the space debris. This "space trucking service" is a planned means of developing a plasma engine for a Mars mission. Someday a 123,000 mph nuclear hydrogen plasma engine may take us to Mars in 39 days.[14]

APPENDICIES

Appendices

A -Unit Conversions

Most engineering calculations are made using metric units of measure. This Appendix presents the conversion from metric units to U. S. English units and U. S. English units to metric units.

Abbreviations			
meter	m	feet	ft
kilometer	km	miles	mi
kilogram	kg	pound	lb
second	s	miles/hour	mph
Units Conversions			
m	=	3.2808	ft
km	=	0.6214	mi
m/s	=	3.2808	ft/s
km/s	=	2236.9363	mph
kg	=	2.2046	lb
Newton	=	0.2248	lbf
ft	=	0.3048	m
mi	=	1.6092	km
ft/s	=	0.3048	m/s
mph	=	0.000447	km/s
lb	=	0.4536	kg
lbf	=	4.4483	Newton

APPENDICIES

B - Longitude Distribution

Longitude Range	Number of Satellites	Major City
180 W - 165 W	4	Adak, Alaska
165 W - 150 W	1	Honolulu, Hawaii
150 W - 135 W	5	Fairbanks, Alaska
135 W -120 W	12	San Francisco, CA
120 W - 105 W	20	Los Angeles, CA
105 W - 90 W	36	Minneapolis, MN
90 W - 75 W	15	Chicago, IL
75 W - 60 W	19	New York, NY
60 W - 45 W	10	Brasilia, Brazil
45 W - 30 W	14	Rio de Janerio, Brazil
30 W - 15 W	12	Dakar, Senegal
15 W - 0	24	London, England
0 - 15 E	22	Paris, France
15 E - 30 E	28	Athens, Greece
30 E - 45 E	19	Moscow, Russia
45 E - 60 E	20	Tehran, Iran
60 E - 75 E	20	Kabul, Afghanistan
75 E - 90 E	24	New Delhi, India
90 E - 105 E	20	Singapore, Singapore
105 E - 120 E	22	Hong Kong, China
120 E - 135 E	12	Shanghai, China
135 E - 150 E	13	Tokyo, Japan
150 E - 165 E	8	Sydney, Australia
165 E - 180 E	7	Christchurch, New Zealand

APPENDICIES

C - Operators/Contractors

U.S. Operators/Owners	Contractor
Civil	
Hawk Institute for Space Science	Hawk Institute for Space Sciences/Pumpkin
University of Miami, RSMAS (loaned by NOAA)	Hughes Aircraft
University of Texas - Austin	University of Texas - Austin
US Air Force Academy	Air Force Academy
US Naval Academy	Midshipmen of U.S. Naval Academy
Civil/Government	
NASA-Ames Research Center/Stanford University	NASA/Ames Research Center
University of Michigan/SRI	University of Michigan
Commercial	
1Worldspace	Alcatel Space Industries
Bigelow Aerospace	Bigelow Aerospace
DigitalGlobe Corporation	Ball Aerospace
DirecTV, Inc.	Boeing Satellite Development Center
Echostar Technologies, LLC	Lockheed Martin Commercial Space Systems
Globalstar	Space Systems/Loral/Alenia Aerospazio
Hughes Network Systems	Boeing Satellite Systems
Intelsat, Ltd.	Aerospatiale
LightSquared	Boeing Satellite Systems
ORBCOMM Inc.	Orbital Sciences Corp.
PanAmSat (Intelsat, Ltd.)	Boeing Satellite Systems
SAT-GE	Alcatel Alenia Space
SES (Societe Europienne des Satellites)	Alcatel Space Industries
SES World Skies (SES [Societe Europienne des Satellites])	Boeing Satellite Systems
Sirius Satellite Radio	Space Systems/Loral
Sirius XM Radio, Inc.	Space Systems/Loral
TerraStar Corporation	Space Systems/Loral
WildBlue Communications	Space Systems/Loral
XM Satellite Radio USA	Boeing Satellite Systems

U.S. Operators/Owners	Contractor
Commercial/Government	
GeoEye	General Dynamics
Government	
Center for Atmospheric Sciences, Hampton University/NASA	Orbital Sciences Corp.
Goddard Space Flight Center (NASA)	Engineering Director, Goddard Space Flight Center
Los Alamos National Labs (LANL)	Surrey Satellite Technologies Ltd.
NASA Earth Science Office	Swales Aerospace
NASA Goddard Space Flight Center	TRW Space and Electronics
NASA Goddard Space Flight Center, Jet Propulsion Laboratory	Orbital Sciences Corp.
NASA/GSFC	General Dynamics
NASA/US Geological Survey	GE Astro Space
National Aeronautics and Space Administration (NASA)	Boeing Satellite Systems
National Oceanographic and Atmospheric Administration (NOAA)	Lockheed Martin Missiles & Space
NOAA (National Oceanographic and Atmospheric Administration)	Boeing Satellite Systems
Government/Civil	
California Institute of Technology/NASA	Orbital Sciences Corp.
Mission and Science Operations Center, UC Berkeley/NASA	University of California, Berkeley
NASA/Applied Physics Laboratory, Johns Hopkins	Applied Physics Laboratory, Johns Hopkins
NASA/Colorado State University	NASA Jet Propulsion Laboratory
Space Sciences Laboratory, UC Berkeley/NASA	Spectrum Astro, Inc.
Government/Commercial	
Iridium Satellite LLC	Motorola Satellite Communications
NASA/SES Americom (SES [Societe Europienne des Satellites])	TRW Defense and Space Systems Group
SES Americom (SES [Societe Europienne des Satellites])	TRW Defense and Space Systems Group

U.S. Operators/Owners	Contractor
Military	
Air Force Research Laboratory	Air Force Research Laboratory/Raytheon
Defense Advanced Research Projects Agency (DARPA)/Navy Research Laboratory (NRL)	Lockheed Martin Missiles and Space
DoD/NOAA	General Electric Astro Space
Missile Defense Agency (MDA)	Northrop Grumman
National Reconnaissance Office (NRO)	Lockheed Martin
Strategic Space Command/Space Surveillance Network	Boeing/Ball Aerospace
US Air Force	General Dynamics
US Navy	TRW, Defense and Space Systems Group
Military/Commercial	
DoD/US Air Force	Boeing Satellite Systems

APPENDICIES

D - Launch Vehicles and Sites

Launch Vehicle	Launch Site
Ariane 44L	Guiana Space Center
Ariane 5	Guiana Space Center
Atlas 2A	Cape Canaveral
Atlas 5	Cape Canaveral
Atlas Centaur	Cape Canaveral
Atlas E	Vandenberg AFB
Delta	Vandenberg AFB
Delta 2	Cape Canaveral
Delta 2	Vandenberg AFB
Delta 2310	Vandenberg AFB
Delta 2914	Cape Canaveral
Delta 4	Cape Canaveral
Long March	Xichang Satellite Launch Center
Minotaur	Wallops Island Flight Facility
Minotaur	Kodiak Launch Complex
Minotaur	Vandenberg AFB
Pegasus	Cape Canaveral
Pegasus	L-1011 Aircraft
Pegasus	Vandenberg AFB
Pegasus	Wallops Island Flight Facility
Proton	Baikonur Cosmodrome
Zenit	Baikonur Cosmodrome
Zenit	Sea Launch (Odyssey)

APPENDICIES

E - Sample Communications Satellites

Name of Satellite	AMC-9
Country of Operator	USA
Operator/Owner	SES (Societe Europienne des Satellites)
Longitude	82.95 W
Launch Mass (lbs.)	10,985
Dry Mass (lbs.)	5,358
Power (watts)	10,000
Date of Launch	6/7/2003
Expected Lifetime	15 yrs.
Contractor	Alcatel Space Industries
Country of Contractor	France
Launch Site	Baikonur Cosmodrome
Launch Vehicle	Proton K
Transponders	24 Ku-band
Service	TV , government, enterprise networks
Coverage	North America
Name of Satellite	Astra 2C
Country of Operator	Luxembourg
Operator/Owner	SES (Societe Europienne des Satellites)
Longitude	28.21 E
Launch Mass (lbs.)	9,760
Dry Mass (lbs.)	5,358
Power (watts)	7,000
Date of Launch	6/16/2001
Expected Lifetime	15 yrs.
Contractor	Boeing Satellite Systems
Country of Contractor	USA
Launch Site	Baikonur Cosmodrome
Launch Vehicle	Proton K
Transponders	32 Ku-band
Service	Direct-to-home; broadcasting, multimedia
Coverage	UK and Republic of Ireland

Name of Satellite	DirecTV-9S
Country of Operator	USA
Operator/Owner	DirecTV, Inc.
Longitude (degrees)	101.08 W
Launch Mass (lbs.)	14,736
Dry Mass (lbs.)	6,334
Power (watts)	13,900
Date of Launch	10/13/2006
Expected Lifetime	15 yrs.
Contractor	Space Systems/Loral
Country of Contractor	USA
Launch Site	Guiana Space Center
Launch Vehicle	Ariane 5 ECA
Transponders	52 Ku-band, 2 Ka-band
Service	DirecTV, voice, video and internet.
Coverage	North America
Name of Satellite	Echostar 7
Country of Operator	USA
Operator/Owner	Echostar Technologies, LLC
Longitude (degrees)	118.85 W
Launch Mass (lbs.)	10,789
Dry Mass (lbs.)	5,894
Power (watts)	13,000
Date of Launch	2/21/2002
Expected Lifetime	12 yrs.
Contractor	Space Systems/Loral
Country of Contractor	USA
Launch Site	Cape Canaveral
Launch Vehicle	Atlas 3B
Transponders	32 Ku-band
Service	DISH Network
Coverage	CONUS, Hawaii, Alaska and Puerto Rico

Name of Satellite	Eutelsat W-5
Country of Operator	Multinational
Operator/Owner	European Telecom. Satellite Consortium
Longitude (degrees)	70.54 E
Launch Mass (lbs.)	8,493
Dry Mass (lbs.)	5,090
Power (watts)	5,900
Date of Launch	11/20/2002
Expected Lifetime	12 yrs.
Contractor	Alcatel Space Industries
Country of Contractor	France
Launch Site	Cape Canaveral
Launch Vehicle	Atlas IIAS
Transponders	24 Ku-band
Service	TV, radio, VSAT services, broadband
Coverage	Western Europe, Middle East, Asia.
Name of Satellite	Intelsat 905
Country of Operator	USA
Operator/Owner	Intelsat, Ltd.
Longitude (degrees)	24.52 W
Launch Mass (lbs.)	12,654
Dry Mass (lbs.)	5,316
Power (watts)	10,000
Date of Launch	6/5/2002
Expected Lifetime	13 yrs.
Contractor	Space Systems/Loral
Country of Contractor	USA
Launch Site	Guiana Space Center
Launch Vehicle	Ariane 44L
Transponders	76 C-band, 22 Ku-band
Service	Business services, direct-to-home TV
Coverage	Atlantic Ocean region.

Name of Satellite	NSS-7
Country of Operator	Netherlands
Operator/Owner	SES (Societe Europienne des Satellites)
Longitude (degrees)	21.00 W
Launch Mass (lbs.)	12,056
Dry Mass (lbs.)	6,698
Power (watts)	3,900
Date of Launch	4/16/2002
Expected Lifetime	14 yrs.
Contractor	Lockheed Martin Commercial Space Systems
Country of Contractor	USA
Launch Site	Guiana Space Center
Launch Vehicle	Ariane 42L
Transponders	36 C-band, 36 Ku-band
Service	Broadcasting, business services;
Coverage	Europe, Africa, the Middle East, the Americas

F - Earth Parameters

Earth parameters including the Standard Universal Constant are essential to orbit calculations.

Earth Parameters				
Equatorial Radius	6378.14	km	3963.19	mi
Polar Radius	6356.80	km	3949.93	mi
Equatorial Circumference	40075.16	km	24901.55	mi
Meridian Circumference	40008.00	km	24859.82	mi
Mass	5.974×10^{24}	kg	1.3170×10^{25}	lb.
Escape Velocity East	10.73	km/s	6.67	mi/s
Escape Velocity West	11.67	km/s	7.25	mi/s
Equatorial Rotational Velocity	1674.40	km/h	1040.42	mph
Axial Tilt	23.40	deg	23.40	deg
Standard Universal Constant	398600.44	$km^3 s^{-2}$	95629.33	$mi^3 s^{-2}$
			1.23936×10^{12}	$mi^3 hr^{-2}$

APPENDICIES

Glossary

antenna beam: The main lobe of an antenna radiation pattern.

apogee: The point in the orbit of an object (like a satellite) orbiting the earth that is at the greatest distance from the center of the earth.

attitude: orientation

attitude control: Control of the angular position and rotation of the satellite.

backhauls: Getting audio and video material to a main distribution point.

bandwidth: A range of radio frequencies used in radio or telecommunications transmission and reception.

bent pipe: The sending back of what goes into the conduit with only amplification and a shift from uplink to downlink frequency.

braking: Halting the rotation of a momentum wheel when it has been returned to its nominal speed.

broadcast feeds: Broadcasting to and from television networks and local affiliate stations (such as program feeds for network and syndicated programming).

bus: The platform part of the satellite that transports and supports the mission payload.

circular orbit: An orbit that always has the same distance from the center of the earth.

dbW: Decibels relative to one watt of power. One watt is equivalent to 0 dBW. 10 dbW – 10 watts, 20 dbW – 100 watts, 70 dbW – 10,000,000 watts.

decibel: A unit used to express relative difference in power or intensity, usually between two acoustic or electric signals, equal to ten times the common logarithm of the ratio of the two levels.

Delta-V: Change in velocity or speed.

deployment: The unfolding of the antennas and solar arrays.

downlink: The signal from the satellite to earth.

earth anomalies: The non-uniform mass distribution of the earth.

earth sensors: A class of sensors that measure the difference in irradiance between the earth and space.

earth's limb: Edge of earth's disc as seen from space.

ecliptic: The earth's orbital plane around the sun.

electromagnetic radiation: Radiation in the form of electromagnetic waves such as gamma rays, X-rays, ultraviolet light, visible light, infrared radiation, microwaves and radio waves.

elliptical orbits: Orbits that resemble an ellipse. A satellite rotates around one of the foci of the ellipse.

equatorial plane: The plane of the equator.

equinox: An equinox occurs twice a year (around 20 March and 22 September), when the tilt of the Earth's axis is inclined neither away from nor towards the Sun.

eccentricity: A measure of the non-circularity of an elliptical orbit, the distance between the foci divided by the length of the major axis.

firing time: The time a thruster fires.

footprint: The geographical area that can be served by a communications satellite.

frequency band: A range of frequencies.

Geostationary Orbit (GEO): A geostationary orbit is a 22,236 miles altitude circular orbit around the earth with an orbital period that matches the earth's sidereal rotation period with zero inclination with respect to the equatorial plane.

GHz: Gigahertz (billion Hz)

hertz (symbol **Hz**): The hertz is the standard unit of frequency defined as the number of cycles per second of a periodic phenomenon.

inclination angle: The angle between two orbit planes.

inclination drift: A gradual change in the inclination angle between the geostationary orbit plane and the satellite's orbit plane caused by gravitational attractions of the sun and moon.

killer satellites: Satellites designed to destroy other satellites.

kHz: Kilohertz (thousand Hz)

latency: The transmission delay of a signal.

launch shroud: a streamlined protective covering used to protect the payload during a rocket-powered launch.

launch vehicles: Rockets that boost satellites into space.

LEO: Low earth orbits are between 100 and 1,240 miles altitude.

lifetime: The operational life of a satellite.

liquid bipropellant: A combination of a fuel and oxidizer.

longitudinal drift: A gradual satellite drift toward a stable point and away from an unstable point.

mass: Weight.

mass flow: The propellant flow in a rocket engine measured in pounds per second.

mass ratio: The number of pounds of propellant it takes to launch one pound of satellite.

Mbps: Megabits per second or a million bits per second and refers to the speed of data transfer.

MEO: Medium earth orbits are between 1,243 and 22,236 miles altitude.

MHz: Megahertz (million Hz)

mission payload: The mission equipment in the satellite.

modulation: Varying the frequency, amplitude, or other characteristics of a radio wave or another carrier wave in order to transmit information.

momentum wheel: A device used to control the attitude of satellites.

Omni antenna: An antenna which radiates radio wave power uniformly in all directions in one plane.

orbit Injection: Applying a velocity maneuver to change from one orbit to another.

orbital slots: Longitudinal segments of the geostationary orbit.

perigee: The point in the orbit of an object (as a satellite) orbiting the earth that is nearest to the center of the earth.

period of the orbit: The time it takes to make one revolution of an orbit.

perturbation: A small change in a physical system, such as position, velocity, angular rotation, or orbital parameters.

photovoltaic: Able to generate a current or voltage when exposed to visible light or other electromagnetic radiation.

pitch axis: A satellite axis that is orthogonal or at a right angle to the roll-yaw (equatorial) plane.

propellant: A substance that is burned to give thrust to a rocket.

rain fade: A noticeable degradation in receptions due to the problems caused by and proportional to the amount of rainfall.

RF: Radio frequency.

reference axes: An orthogonal reference frame related to the satellite axes.

repeater: A group of transponders that share the antenna and receiver.

roll axis: A satellite axis that points along the direction of travel.

satellite clusters: Closely grouped satellites in an orbital slot.

satellite configurations: Satellites are under a shroud during launch, the solar arrays are stowed or folded during transfer to an operational orbit, and the solar arrays are deployed during operation.

satellite payload: The satellite launched by a launch vehicle.

semi major axis: The major axis of an ellipse is its longest diameter with its ends being at the widest points of the shape. The semi major axis is one half of the major axis.

sidereal day: The length of time which passes between a given "fixed" star in the sky crossing a given projected meridian (line of longitude). The sidereal day is 23 h 56 m 4.1 s, slightly shorter than the solar day because the Earth's orbital motion about the Sun means the Earth has to rotate slightly more than one turn with respect to the "fixed" stars in order to reach the same Earth Sun orientation.

signal relay stations: Land stations that relay microwave signals.

skin effect: When electromagnetic radiation impinges upon a conductor, it couples to the conductor, travels along it, and induces an electric current on the surface of that conductor by exciting the electrons of the conducting material.

snow fade: A loss of reception due snow.

solar array: Assemblies of solar panels.

solar cell: A solid state electrical device that converts the energy of light directly into electricity by the photovoltaic effect

solar modules: Assemblies of solar cells.

solar panels: Assemblies of solar modules.

solstice: A solstice is an astronomical event that happens twice each year when the Sun reaches its highest position in the sky as seen from the North or South Pole.

specific impulse: A measure of the fuel efficiency of a rocket, expressed as the number of pounds of thrust produced per pound of propellant used per second.

speed maneuvers: A change in speed to go from one orbit to another.

stages: The number of rocket engines in a launch vehicle.

standard universal constant: The constant of proportionality of the attractive force between two bodies and the product of their masses, and inversely proportional to the square of the distance between them.

station: The operational position of a geostationary satellite in a geostationary orbit at an assigned longitude over the equator.

station-keeping: Keeping the satellite in the right orbit so it stays on its assigned longitude above the equator.

stowed: A satellite configuration where the antennas and solar arrays are folded close to the satellite body.

subsystems: The next lower level of a system.

sun sensor: A sensor that detects the direction to the sun.

surface circular satellite: has a semi-major axis equal to the radius of the earth

telemetry: Transmitted information to and from a satellite.

THz: Terahertz (trillion Hz)

terrestrial: Of or relating to the earth.

three axis stabilized satellite: A satellite where stabilization is achieved by controlling the rotation of the satellite about all three axes.

thrust: The reactive force of expelled gases generated by a rocket measured in pounds-feet.

torque: Torque is the tendency of a force to rotate an object about an axis.

transfer orbit: An orbit that transfers a satellite from one orbit to another.

transponder: An automatic device that receives, amplifies, and retransmits a signal on a different frequency. Transponders are satellites communications channels.

Traveling-Wave Tube (TWT): An electronic device used to amplify radio frequency signals to high power.

Traveling-Wave Tube Amplifier (TWTA): A RF power amplifier.

uplink: The signal sent from earth to the satellite.

wavelength: The distance between two points on adjacent waves that have the same phase, e.g. the distance between two consecutive peaks or trough.

yaw axis: A satellite axis that points to the center of the earth.

References

[1]Braeuing, Robert A. "Rocket and Space Technology"
http://www.braeunig.us/space/index_top.htm

[2]"UCS Satellite Database". Union of Concerned Scientists. 1 September 2011
http://www.ucsusa.org/nuclear_weapons_and_global_security/space_weapons/technical_issues/ucs-satellite-database.html

[3] Roberts, Lawrence D. "A Lost Connection: Geostationary Satellite Networks and the International Telecommunication Union."
http://www.law.berkeley.edu/journals/btlj/articles/vol15/roberts/roberts.html

[4] Todd, Laura and Tom Bowling "Debris Mitigation in Geostationary Earth Orbit." 2004
http://naca.central.cranfield.ac.uk/dcsss/2004/C23_ToddEDITfinal.pdf

[5] "Space Junk Threat Will Grow for Astronauts and Satellites." Foxnews.com. April 06, 2011 http://www.foxnews.com/scitech/2011/04/06/space-junk-threat-grow-astronauts-satellites/

[6] Iannotta, Becky and Tarig Malik "U.S. Satellite Destroyed in Space Collision" space.com 2009
http://www.space.com/5542-satellite-destroyed-space-collision.html

[7] "Intelsat." 10 October 2011 http://www.intelsat.com/

[8] Green, David "Satellite Costs" Satellite Evolution http://www.satellite-evolution.com/PDF%20Files/JulyAugust%20Issue/LMCS.pdf

[9] "Expendable Space Vehicles" Space Tech
http://www.spaceandtech.com/spacedata/elvs/elvs.shtml

[10] "Earth Sensors." Sodern.
http://www.sodern.com/site/FO/scripts/siteFO_contenu.php?mode=&noeu_id=61&lang=EN

[11] "Fine Sun Sensor." SSBV Space & Ground Systems UK
http://www.satserv.co.uk/Products_ADCS_Sensors_FineSunSensor.html

[12] "Propulsion Subsystem." GEOS I-M Databook, Revision 1, 1996
http://rsd.gsfc.nasa.gov/goes/text/goes.databook.html

[13] "NASA Plans to Refuel Mock Satellite at the Space Station." Space News.2
April 2010 http://www.spacenews.com/civil/100402-nasa-plans-refuel-mock-satellite.html

[14] "Mars In 39 Days", by Sam Howe Verhovek, Popular Science Magazine,
November 2010, Volume 277 #5

Index

About the Author

C. Robert (Bob) Welti's career spans over fifty years in aerospace and software systems. He is in the space business since Sputnik. His experience includes: Engineer, Project Manager, Program Manager, Director, and Consultant on military and commercial satellite, telecommunications, and missile programs. He worked on reconnaissance, interceptor, and communications satellites; the space shuttle; and the Minuteman ballistic missile.

He managed software development and maintenance for the U.S. Air Forces' Satellite Control Facility. At the Aerospace Corporation, he performed system engineering and proposal evaluation on many U.S. Air Force satellite programs. He also evaluated many commercial companies for their software development and maintenance capabilities. He taught classes on project management, astrodynamics, digital electronics, engineering mathematics, and computer system design.

Bob earned his Ph.D. in Engineering from UCLA with a major in the self-developed field of "Space Vehicle Dynamics." His minor fields were Computers and Management. His dissertation is "Recursive Navigation and Guidance for the Midcourse Phase of Orbital Rendezvous." He has a MSEE from USC, a BSEE from Columbia University and a BA from Hofstra University. Also for two years, he learned about and gained hands-on experience in Marine Engineering while a midshipman at the United States Merchant Marine Academy (USMMA).

He now enjoys retirement living on a golf course with his wife Kem in the foothills of the California Gold Country between Sacramento and Lake Tahoe. Besides golf he enjoys wood working, oil painting, and spending time with his family.

Printed in the United States
By Bookmasters